Principles of Applied Optics

Principles of Applied Optics

Partha P. Banerjee
Syracuse University

Ting-Chung Poon
Virginia Polytechnic Institute
and State University

Aksen
Associates
Incorporated
Publishers

IRWIN

Homewood, IL 60430
Boston, MA 02116

Cover and text designer: Harold Pattek
Compositor: Science Typographers, Inc.
Typeface: Times Roman with Vega
Printer: R. R. Donnelley and Sons

Library of Congress Cataloging-in-Publication Data

Banerjee, Partha P.
    Principles of applied optics/Partha P. Banerjee, Ting-Chung Poon.
        p.   cm.—(The Aksen Associates series in electrical and computer engineering)
    Includes index.
    ISBN 0-256-08860-8
    1. Optics.   I. Poon, Ting-Chung.   II. Title.   III. Series.
QC355.2.B36   1991
621.36—dc20   90–1183

Printed in the United States of America

1 2 3 4 5 6 7 8 9 0 DOC 7 6 5 4 3 2 1 0

About the Authors

Partha P. Banerjee is associate professor of electrical and computer engineering at Syracuse University in Syracuse, New York. Dr. Banerjee received his B.Tech. degree at the Indian Institute of Technology and his M.S. and Ph.D. degrees in electrical and computer engineering at the University of Iowa. He has published over 50 technical papers in a variety of areas, including nonlinear wave phenomena, nonlinear optics and acoustics, optical information processing, acoustooptics, optical computing, and electromagnetics. He has also given presentations on nonlinear optics and acoustooptics internationally. Dr. Banerjee consults for industry and teaches courses on optics for professionals. Dr. Banerjee is a recipient of an NSF Presidential Young Investigator award. He is a member of the Optical Society of America and the Society of Industrial and Applied Mathematics and is a senior member of the Institute of Electrical and Electronics Engineers.

Ting-Chung Poon is associate professor of electrical engineering at Virginia Polytechnic Institute and State University in Blacksburg, Virginia. Dr. Poon received his B.A. degree in physics and mathematics and his M.S. and Ph.D. degrees in electrical and computer engineering at the University of Iowa. He has published over 40 technical papers and has given lectures and seminars on applied optics internationally. His areas of specialization are acoustooptics, diffractive optics, hybrid (optical–electronic) image processing, and optical scanning holography. He is a member of the Optical Society of America and the Society of Photo-optical Instrumentation Engineers and is a senior member of the Institute of Electrical and Electronics Engineers.

Contents

Preface

Optics is a growing field. This activity reaches into most scientific and engineering disciplines, and there is currently a definite need in academia as well as in industry for qualified individuals who are trained in the optical sciences. In writing this book, we have directed our efforts to familiarizing seniors and first-year graduate students in engineering and physics to principles and applications in modern optics. Readers should realize that this is not, therefore, an exclusive treatise on optics. Neither is it supposed to be confined to a specialized area such as Fourier optics or optical electronics. It is our hope that this book will stimulate the readers' general interest as well as provide them with a broad background in optics.

The principal objective of the book is to provide a comprehensive approach to the broad field of optics. The main topics covered are geometrical and physical optics, the Fourier-transforming property of lenses, and optical information processing (including complex spatial filtering and holography), acoustooptic and electrooptic effects, principles of lasers and photodetectors, and some topics from nonlinear optics (such as harmonic generation, phase conjugation, and optical bistability). The material in this book is intended for a two-semester sequence of courses with some supplementary notes from the instructor. Chapters 1 through 5 are suitable for a course in optical information processing, and Chapters 6 through 8, with a brief review of Chapters 3 through 5, are appropriate for a course in optoelectronics. Parts of the text were presented in a first-year graduate-level course in optical information processing for two years and in a graduate-level course on nonlinear optics at Syracuse University, and in a senior-level course at Virginia Polytechnic Institute. We recommend previous exposure to basic electromagnetics and communication theory, as well as fundamentals of solid-state devices, for understanding most of the material that is covered in the book.

Chapter 1 introduces mathematical conventions and tools to be used in the subsequent chapters. The chapters on geometrical (Chapter 2) and physical (Chapter 3) optics provide readers with necessary background for the chapters that follow. We point out that in the chapter on geometrical optics, we employ a matrix formalism to describe optical image formation. The advantage of this is that any ray can be tracked during its propagation through the optical system by successive matrix multiplications, which can be easily programmed on a digital computer. This representation of geometrical optics is elegant and powerful and is widely used in the design of optical elements.

In Chapter 3, we start with the discussion of linear wave propagation and develop the eikonal equations for ray propagation. We want to point out that our treatment of diffraction is different from the conventional approach in the sense that we derive the diffraction formula directly from the wave equation using Fourier transforms, without going through the Kirchhoff or Rayleigh–Sommerfeld formulations. In this way, we not only avoid complicated mathematical manipulations involving Green's functions, but also give readers a direct systems-oriented insight into the diffraction phenomenon, namely, the transfer function of propagation, and the corresponding impulse response. In addition to covering wave propagation in homogeneous media, we also discuss wave propagation in inhomogeneous media, where we take beam propagation in graded-index optical fibers as an example.

In Chapter 4, we discuss tools and techniques for optical information processing, including the transforming and image properties of lenses, using wave optics. We also introduce the concept of impulse response and transfer function in optical imaging systems and discuss how different image processing operations can be achieved by modifying the impulse responses via aperture functions.

In Chapter 5, we cover the principles of holography and discuss the construction of optical holograms and complex spatial filters that are the essential elements in optical systems with applications to pattern recognition and image processing. Owing to the similarities in the techniques of construction and use of holograms and spatial filters, we find it appropriate to treat both topics simultaneously.

Chapter 6 discusses acoustooptic and electrooptic effects. The ability to modulate light waves by electrical signals either through the acoustooptic or electrooptic effect provides a powerful means for optically processing information. In this chapter, wave propagation in anisotropic media (such as crystals) is also discussed. We purposely postponed this discussion to Chapter 6 instead of presenting it in Chapter 3 to keep the text as simple and readable as possible.

No treatment of modern optics would be complete without mentioning the sources of coherent light and detectors. These are discussed in Chapter 7, where we cover simplified laser theory and laser optics. We discuss how, in a two-level atomic system, the effect of population inversion enhances stimulated emission leading to coherent radiation, which can be amplified in a resonator. Using diffraction theory developed in Chapter 3, we show that the output optical beam profile is, in general, Hermitian–Gaussian. Different photodetector circuits are also presented, along with their performance in terms of the signal-to-noise ratio.

Finally, in Chapter 8, we cover some relatively modern developments in nonlinear optics, such as harmonic generation, self-refraction, bistability, phase conjugation, and soliton propagation, which we feel readers should be exposed to at this course level. Once again, our treatment is different from conventional approaches in the sense that we expose readers to the concepts and essential mathematical models while avoiding, for the most part, any use of tensors, which, we feel, often eclipses the basic physics.

A comment concerning units and notation is in order. Throughout the book, we use the MKS system of units and the engineering convention for wave propagation, to be made precise in the text.

We are indebted to the individuals who typed parts of the manuscript. These include Pat White and Sandra Howell of Virginia Tech and Bert Fancher of Syracuse University. We would also like to express our gratitude to Dr. Adrian Korpel of the University of Iowa for instilling in us the spirit of optics and without whose guidance, we would not be where we are. We would also take this opportunity to thank Dr. Paul Prucnal of Princeton University, Dr.

Lloyd Hillman of the University of Alabama, and Dr. Doreen Weinberger of the University of Michigan who reviewed the manuscript and offered numerous helpful suggestions. Thanks are also due to Dr. Guanghui Cab of Virginia Commonwealth University who read parts of the manuscript. We would also like to acknowledge the support and encouragement from our respective Department Chairmen during the preparation of the manuscript. P. P. Banerjee would like to gratefully acknowledge the whole-hearted support from Noriko Tsuchihashi since the inception of this project. T. − C. Poon would like to thank his wife, Eliza Lau, and his children, Christina and Justine, for their encouragement, patience, and love. Last, but not least, we are greatly indebted to our parents, whose moral encouragement and personal sacrifice have enabled us to complete this project.

Chapter 1 Mathematical Preliminaries

The purpose of this chapter is to present mathematical basics and some nomenclature that we will use throughout the book. First, we introduce the Fourier transform in two dimensions and discuss some of its properties. We then review the properties of linear systems, including the concept of the transfer function and impulse response. Thereafter, we introduce the Einstein convention for representing tensors, which we will use in connection with electrooptics and nonlinear optics. Finally, we present some notation, including definitions of the real field or wavefunction and the corresponding complex phasor and complex envelope, that we will use time and again throughout the book.

1.1 The Two-Dimensional Fourier Transform and Its Properties

The *Fourier transform* in one dimension of a square-integrable function $f(t)$ is defined as

$$F(\omega) = \int_{-\infty}^{\infty} f(t)\exp[-j\omega t]\, dt \triangleq \mathscr{F}_t\{f(t)\}. \qquad (1.1\text{-}1)$$

The inverse transform is

$$f(t) = \frac{1}{2\pi}\int_{-\infty}^{\infty} F(\omega)\exp[j\omega t]\, d\omega \triangleq \mathscr{F}_t^{-1}\{F(\omega)\}. \qquad (1.1\text{-}2)$$

Mathematical Preliminaries

In Eqs. (1.1-1) and (1.1-2), t denotes time and ω the temporal frequency. Hence $F(\omega)$ is the *temporal Fourier transform* of $f(t)$. We can easily check that

$$\mathscr{F}_t^{-1}\{\mathscr{F}_t\{f(t)\}\} = f(t), \qquad \mathscr{F}_t\{\mathscr{F}_t^{-1}\{F(\omega)\}\} = F(\omega). \quad (1.1\text{-}3)$$

Square integrability, that is,

$$\int_{-\infty}^{\infty} |f(t)|^2 \, dt < \infty, \qquad\qquad (1.1\text{-}4)$$

is a necessary condition for the existence of the Fourier transform, although Fourier transforms of non–square-integrable "functions" or distributions often exist.

A list of Fourier-transform pairs of some commonly used time functions is shown in Table 1.1.

In image processing, we are often interested in the Fourier transform of a two-dimensional (2-D) function $f(x, y)$, where x and y are spatial coordinates. The transform variables are now k_x, k_y and are called *spatial frequencies*. Analogous to Eqs. (1.1-1) and

Table 1.1 Fourier-Transform Pairs of Commonly Used Time Functions

$f(t)$	$F(\omega)$
1. Delta function $\delta(t)$	
$\displaystyle \delta(t) = \frac{1}{2\pi} \int_{-\infty}^{\infty} e^{\pm j\omega t} \, d\omega$	1
2. 1	$2\pi\delta(\omega)$
3. Unit step function $U(t)$	
$U(t) = \begin{cases} 1, & t > 0, \\ 0, & t < 0 \end{cases}$	$\pi\delta(\omega) + \dfrac{1}{j\omega}$
4. $\cos \omega_0 t$, ω_0 constant	$\pi[\delta(\omega - \omega_0) + \delta(\omega + \omega_0)]$
5. $\sin \omega_0 t$	$-j\pi[\delta(\omega - \omega_0) - \delta(\omega + \omega_0)]$
6. $e^{-at}U(t)$, $a > 0$, real	$\dfrac{1}{a + j\omega}$
7. $e^{-at}(\sin \omega_0 t)U(t)$	$\dfrac{\omega_0}{(a + j\omega)^2 + \omega_0^2}$
8. $e^{-at}(\cos \omega_0 t)U(t)$	$\dfrac{a + j\omega}{(a + j\omega)^2 + \omega_0^2}$

(1.1-2), we define the *spatial Fourier transform* pair as

$$F(k_x, k_y) = \int_{-\infty}^{\infty} \int_{-\infty}^{\infty} f(x, y) \exp\left[j(k_x x + k_y y) \right] dx\, dy$$

$$= \mathscr{F}_{xy}\{ f(x, y) \},$$ (1.1-5)

$$f(x, y) = \frac{1}{4\pi^2} \int_{-\infty}^{\infty} \int_{-\infty}^{\infty} F(k_x, k_y) \exp\left[-j(k_x x + k_y y) \right] dk_x\, dk_y$$

$$= \mathscr{F}_{xy}^{-1}\{ F(k_x, k_y) \}.$$ (1.1-6)

The square-integrability condition in two dimensions reads

$$\int_{-\infty}^{\infty} \int_{-\infty}^{\infty} |f(x, y)|^2 \, dx\, dy < \infty$$ (1.1-7)

and is a necessary condition for the existence of the Fourier transform. Also, we can readily establish that

$$\mathscr{F}_{xy}^{-1}\{ \mathscr{F}_{xy}\{ f(x, y) \} \} = f(x, y),$$

$$\mathscr{F}_{xy}\{ \mathscr{F}_{xy}^{-1}\{ F(k_x, k_y) \} \} = F(k_x, k_y).$$ (1.1-8)

We remark that our definitions of $\mathscr{F}_{xy}$ and $\mathscr{F}_{xy}^{-1}$ are different from the definitions of $\mathscr{F}_t$ and $\mathscr{F}_t^{-1}$ in that, for the former, we use $+jk_x x + jk_y y$ and $-jk_x x - jk_y y$ as the exponents for the forward and inverse transforms, respectively, whereas $-j\omega t$ and $+j\omega t$ serve as the exponents in the latter. This is purposely done to be consistent with the engineering convention for a travelling wave. In this convention, $\psi(z, t) = \text{Re}[\, A \exp j(\omega t - kz)]$ generically denotes a wave travelling in the $+z$ direction, having a temporal frequency ω and a propagation constant k. More on this will appear in Chapter 3, following a discussion of Maxwell's equations and the wave equation. We can represent an arbitrary signal $\psi(x, y, z, t)$ as

$$\psi(x, y, z, t) = \frac{1}{2\pi} \int_{-\infty}^{\infty} \Psi(x, y, z; \omega) \exp\left[j(\omega t - kz) \right] d\omega, \quad (1.1\text{-}9)$$

that is, as a collection of plane waves with spectral amplitudes $\Psi(x, y, z; \omega)$. Note that Eq. (1.1-9) is consistent with the definition of the inverse Fourier transform $\mathscr{F}_t^{-1}$, as defined in Eq. (1.1-2). Also observe that if the decomposition of $\psi(x, y, z, t)$ were made in terms of spectral amplitudes such as $\Psi(k_x, k_y; z, t)$, the logical representation for ψ would be

$$\psi(x, y, z, t) = \frac{1}{4\pi^2} \int_{-\infty}^{\infty} \int_{-\infty}^{\infty} \Psi(k_x, k_y; z, t)$$

$$\times \exp\left[j(\omega t - k_x x - k_y y - k_z z) \right] dk_x \, dk_y, \quad (1.1\text{-}10)$$

with $k_z^2 + k_x^2 + k_y^2 = k^2$, which is consistent with the definition of the inverse transform $\mathscr{F}_{x,y}^{-1}$, as defined in Eq. (1.1-6). We remark that k and ω are related to each other via the *dispersion relation* $\omega = W(k)$ or $k = K(\omega)$. For linear wave propagation in vacuum, $\omega = ck$, where c is the velocity of the wave in vacuum. The preceding concepts will become clearer as we work our way through the text.

In Table 1.2, we summarize some of the properties of the two-dimensional Fourier transform, defined in Eqs. (1.1-5) and (1.1-6). More properties will appear in Section 1.2. Some useful Fourier-transform pairs are listed in Table 1.3.

Table 1.2 Some Properties of the Two-Dimensional Fourier Transform

Function in (x, y)	Fourier Transform in (k_x, k_y)
1. $f(x, y)$	$F(k_x, k_y)$
2. $f(x - x_0, y - y_0)$; x_0, y_0 real constants	$F(k_x, k_y)\exp\left[+jk_x x_0 + jk_y y_0 \right]$
3. $f(ax, by)$; a, b complex constants	$\dfrac{1}{|ab|}F\left(\dfrac{k_x}{a}, \dfrac{k_y}{b} \right)$
4. $f^*(x, y)$	$F^*(-k_x, -k_y)$
5. $f(x, y)\exp\left[-jk_{x_0} x - jk_{y_0} y \right]$; k_{x_0}, k_{y_0} real	$F(k_x - k_{x_0}, k_y - k_{y_0})$
6. $F(x, y)$	$4\pi^2 f(-k_x, -k_y)$
7. $\dfrac{\partial}{\partial x} f(x, y)$	$-jk_x F(k_x, k_y)$

Table 1.3 Useful Fourier-Transform Pairs

$f(x, y)$	$F(k_x, k_y)$
1. Delta function, $\delta(x, y)$ $$\delta(x, y) = \frac{1}{4\pi^2} \iint\limits_{-\infty}^{\infty} e^{\pm jk_x x \pm jk_y y}\, dk_x\, dk_y$$	1
2. 1	$4\pi^2\delta(k_x, k_y) = 4\pi^2\delta(k_x)\delta(k_y)$
3. Rectangle function $$\text{rect}(x, y) = \text{rect}(x)\text{rect}(y),$$ where $\text{rect}(x) = \begin{cases} 1, & |x| \le \frac{1}{2} \\ 0, & \text{otherwise} \end{cases}$	Sinc function $$\text{sinc}\left(\frac{k_x}{2\pi}, \frac{k_y}{2\pi}\right) = \text{sinc}\left(\frac{k_x}{2\pi}\right)\text{sinc}\left(\frac{k_y}{2\pi}\right),$$ where $\text{sinc}(k_x) = \dfrac{\sin \pi k_x}{\pi k_x}$
4. Gaussian function $$\exp\left[-\alpha(x^2 + y^2)\right]$$	Gaussian function $$\frac{\pi}{\alpha}\exp\left[-\frac{k_x^2 + k_y^2}{4\alpha}\right]$$
5. $\exp\left[-jk_{x_0}x - jk_{y_0}y\right]$	$4\pi^2\delta(k_x - k_{x_0}, k_y - k_{y_0})$
6. Triangle function $$\Lambda(x, y) = \Lambda(x)\Lambda(y),$$ where $\Lambda(x) = \begin{cases} 1 - |x|, & |x| \le 1, \\ 0, & \text{otherwise} \end{cases}$	Sinc-squared function $$\text{sinc}^2\left(\frac{k_x}{2\pi}\right)\text{sinc}^2\left(\frac{k_y}{2\pi}\right)$$
7. Comb function $$\text{comb}\left(\frac{x}{x_0}, \frac{y}{y_0}\right) = \text{comb}\left(\frac{x}{x_0}\right)\text{comb}\left(\frac{y}{y_0}\right),$$ where $\text{comb}\left(\dfrac{x}{x_0}\right) = \displaystyle\sum_{n=-\infty}^{\infty} \delta(x - nx_0)$	Comb function $$k_{x_0}k_{y_0}\text{comb}\left(\frac{k_x}{k_{x_0}}, \frac{k_y}{k_{y_0}}\right),$$ $$k_{x_0} = \frac{2\pi}{x_0},\ k_{y_0} = \frac{2\pi}{y_0},$$ where $\text{comb}\left(\dfrac{k_x}{k_{x_0}}\right) = \displaystyle\sum_{n=-\infty}^{\infty} \delta(k_x - nk_{x_0})$

Example 1.1

We will derive the two-dimensional Fourier transform of the function $f((x - x_0)/\alpha, (y - y_0)/\beta)$, with α, β real and positive, assuming $f(x, y)$ is Fourier-transformable and

$$f(x, y) \underset{\mathscr{F}_{xy}^{-1}}{\overset{\mathscr{F}_{xy}}{\longleftrightarrow}} F(k_x, k_y).$$

From first principles,

$$\mathscr{F}_{xy}\left\{ f\left(\frac{x - x_0}{\alpha}, \frac{y - y_0}{\beta} \right) \right\}$$

$$= \int_{-\infty}^{\infty} \int_{-\infty}^{\infty} f\left(\frac{x - x_0}{\alpha}, \frac{y - y_0}{\beta} \right) \exp\left[j(k_x x + k_y y) \right] dx\, dy.$$

Substituting $X = (x - x_0)/\alpha$ and $Y = (y - y_0)/\beta$, we recast the preceding integral into the form

$$\exp\left[j(k_x x_0 + k_y y_0) \right] \int_{-\infty}^{\infty} \int_{-\infty}^{\infty} f(X, Y) \exp\left(j[\alpha k_x X + \beta k_y Y] \right) dX\, dY$$

$$= \alpha\beta \exp\left[j(k_x x_0 + k_y y_0) \right] F(\alpha k_x, \beta k_y).$$

Alternatively, we can derive this answer by referring to properties of the Fourier transform in Table 1.2. First, we rewrite the given function as $g(x - x_0, y - y_0)$. Using property 2 from the table, its Fourier transform is $\exp[j(k_x x_0 + k_y y_0)]G(k_x, k_y)$, where $G(k_x, k_y)$ is the Fourier transform of $g(x, y)$. Now, $g(x, y) = f(x/\alpha, y/\beta)$; hence, using property 3 from the table,

$$G(k_x, k_y) = \alpha\beta \exp\left[j(k_x x_0 + k_y y_0) \right] F(\alpha k_x, \beta k_y),$$

identical to the result derived from first principles.

1.2 Linear Systems
A *system* is the mapping of an input or set of inputs into an output or set of outputs. A convenient representation of a system is a mathematical operator. For instance, for a single-input–single-output system,

$$f_o(x, y) = P_{xy}\{f_i(x, y)\}, \tag{1.2-1}$$

where f_i and f_o represent the input and the output, respectively, and where P_{xy} is the operator.

Linear Systems

A system is *linear* if for all complex constants a and b,

$$P_{xy}\{af_{i_1}(x, y) + bf_{i_2}(x, y)\} = aP_{xy}\{f_{i_1}(x, y)\} + bP_{xy}\{f_{i_2}(x, y)\},$$

$$(1.2\text{-}2)$$

that is, the overall output is a weighted sum of the outputs due to inputs f_{i_1} and f_{i_2}. This feature is particularly useful in constructing the output for a given input, knowing the output for an elementary input like the delta function.

For a delta function input of the form $\delta(x - x', y - y')$, the output $P_{xy}\{\delta(x - x', y - y')\} \triangleq \tilde{h}(x, y, x', y')$ is called the *impulse response* of the linear system. Using the *sifting property* of the delta function, we know that an arbitrary function $f_i(x, y)$ can be represented as

$$f_i(x, y) = \int_{-\infty}^{\infty}\int_{-\infty}^{\infty} f_i(x', y')\delta(x - x', y - y') \, dx' \, dy', \quad (1.2\text{-}3)$$

that is, $f(x, y)$ can be regarded as a linear combination of weighted and displaced delta functions. We can then write the output $f_o(x, y)$ of the linear system as

$$f_o(x, y) = P_{xy}\{f_i(x, y)\}$$

$$= \int_{-\infty}^{\infty}\int_{-\infty}^{\infty} f_i(x', y')P_{xy}\{\delta(x - x', y - y')\} \, dx' \, dy'$$

$$= \int_{-\infty}^{\infty}\int_{-\infty}^{\infty} f_i(x', y')\tilde{h}(x, y, x', y') \, dx' \, dy'. \quad (1.2\text{-}4)$$

Now, a linear system is called *space-invariant* if the impulse response $\tilde{h}(x, y, x', y')$ only depends on $x - x', y - y'$, that is,

$$\tilde{h}(x, y, x', y') = h(x - x', y - y'). \quad (1.2\text{-}5)$$

Thus, for linear space-invariant systems, the output $f_o(x, y)$ from Eq. (1.2-4) can be rewritten as

$$f_o(x, y) = \int_{-\infty}^{\infty}\int_{-\infty}^{\infty} f_i(x', y')h(x - x', y - y') \, dx' \, dy'. \quad (1.2\text{-}6)$$

Defining the *convolution* of two functions $g_1(t)$ and $g_2(t)$ as

$$g(t) = g_1(t) * g_2(t) \triangleq \int_{-\infty}^{\infty} g_1(t') g_2(t - t') \, dt', \qquad (1.2\text{-}7)$$

the two-dimensional extension of this for two functions $g_1(x, y)$ and $g_2(x, y)$ is

$$g(x, y) = g_1(x, y) * g_2(x, y)$$

$$\triangleq \int_{-\infty}^{\infty} \int_{-\infty}^{\infty} g_1(x', y') g_2(x - x', y - y') \, dx' \, dy'. \quad (1.2\text{-}8)$$

Referring back to Eq. (1.2-6), we then recognize that the output $f_o(x, y)$ of a linear space-invariant system is a convolution between the input $f_i(x, y)$ and the impulse response $h(x, y)$ of the system, that is,

$$f_o(x, y) = f_i(x, y) * h(x, y) \qquad (1.2\text{-}9a)$$

$$= h(x, y) * f_i(x, y). \qquad (1.2\text{-}9b)$$

We can readily prove Eq. (1.2-9b) from Eq. (1.2-9a) by explicitly writing the convolution integral and making a change of variables.

It is instructive, at this point, to find the Fourier transform $G(k_x, k_y)$ of $g(x, y)$ in Eq. (1.2-8) in terms of the transforms $G_1(k_x, k_y)$ and $G_2(k_x, k_y)$ of $g_1(x, y)$ and $g_2(x, y)$, respectively.

$$G(k_x, k_y) = \mathscr{F}_{xy}\{g(x, y)\} = \mathscr{F}_{xy}\{g_1(x, y) * g_2(x, y)\}$$

$$= \int_{-\infty}^{\infty} \int_{-\infty}^{\infty} \exp\left[j(k_x x + k_y y)\right]$$

$$\times \left[\int_{-\infty}^{\infty} \int_{-\infty}^{\infty} g_1(x', y') g_2(x - x', y - y')\right] dx' \, dy' \, dx \, dy$$

$$= \int_{-\infty}^{\infty} \int_{-\infty}^{\infty} g_1(x', y') \left[\int_{-\infty}^{\infty} \int_{-\infty}^{\infty} g_2(x - x', y - y')\right.$$

$$\left. \times \exp\left[j(k_x x + k_y y)\right] dx \, dy\right] dx' \, dy'.$$

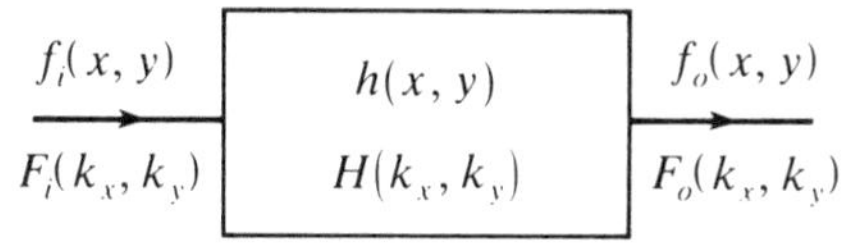

Figure 1.1 Block-diagrammatic representation of a linear space-invariant system.

Using property 2 in Table 1.2, we can reexpress this as

$$G(k_x, k_y) = G_2(k_x, k_y) \int_{-\infty}^{\infty} \int_{-\infty}^{\infty} g_1(x', y') \exp\left[j(k_x x' + k_y y') \right] dx' dy',$$

or

$$G(k_x, k_y) = G_1(k_x, k_y) G_2(k_x, k_y), \qquad (1.2\text{-}10)$$

that is, the Fourier transform of a convolution of two functions is a product of their Fourier transforms. Incorporating this property into Eqs. (1.2-9a) and (1.2-9b), we have

$$F_o(k_x, k_y) = H(k_x, k_y) F_i(k_x, k_y), \qquad (1.2\text{-}11)$$

where $F_o(k_x, k_y)$, $F_i(k_x, k_y)$, and $H(k_x, k_y)$ are the Fourier transforms of $f_o(x, y)$, $f_i(x, y)$, and $h(x, y)$, respectively. $H(k_x, k_y)$ is called the *transfer function* of the system and indicates the behavior of the system in the spatial frequency domain (see Figure 1.1).

Example 1.2

The *correlation* of two functions $s_1(x, y)$ and $s_2(x, y)$ is defined as

$$s(x, y) = s_1(x, y) \circledast s_2(x, y)$$

$$= \int_{-\infty}^{\infty} \int_{-\infty}^{\infty} s_1^*(x', y') s_2(x + x', y + y') dx' dy'. \quad (1.2\text{-}12)$$

We will relate the Fourier transform $S(k_x, k_y)$ of $s(x, y)$ to the

Fourier transforms $S_1(k_x, k_y)$ and $S_2(k_x, k_y)$ of $s_1(x, y)$ and $s_2(x, y)$, respectively. Putting $x'' = -x'$ and $y'' = -y'$ in Eq. (1.2-12), we have

$$s(x, y) = \int_{-\infty}^{\infty} \int_{-\infty}^{\infty} s_1^*(-x'', -y'') s_2(x - x'', y - y'')\, dx''\, dy''$$

$$= s_1^*(-x, -y) * s_2(x, y).$$

Hence, $S(k_x, k_y) = \mathscr{F}_{xy}\{s_1^*(-x, -y)\} S_2(k_x, k_y)$. Now, using properties 3 and 4 of Fourier transforms in Table 1.2, it follows that $\mathscr{F}_{xy}\{s_1^*(-x, -y)\} = S_1^*(k_x, k_y)$. Thus,

$$S(k_x, k_y) = S_1^*(k_x, k_y) S_2(k_x, k_y). \tag{1.2-13}$$

We will extensively use the concepts of convolution and the impulse response and transfer function in various contexts, for instance, in describing the effects of diffraction on wave propagation and in characterizing imaging systems. Also, properties of correlation will be useful in the discussion of holography and spatial filtering.

1.3 Tensors and the Einstein Convention

Tensors are important in many areas of physics and engineering, including general relativity, crystal physics, electromagnetics, electrooptics, and, more recently, nonlinear optics. For instance, as we will see in Chapter 3, the vector electric displacement **D** is related to the vector electric field **E** in an isotropic medium by

$$\mathbf{D} = \epsilon \mathbf{E}, \tag{1.3-1}$$

where ϵ is a scalar quantity. Assuming a Cartesian coordinate system, Eq. (1.3-1) can be written out in its entirety as

$$\mathbf{D} = \begin{pmatrix} D_1 \\ D_2 \\ D_3 \end{pmatrix} = \epsilon \begin{pmatrix} E_1 \\ E_2 \\ E_3 \end{pmatrix} = \epsilon \begin{bmatrix} 1 & 0 & 0 \\ 0 & 1 & 0 \\ 0 & 0 & 1 \end{bmatrix} \begin{bmatrix} E_1 \\ E_2 \\ E_3 \end{bmatrix} = \epsilon I \begin{bmatrix} E_1 \\ E_2 \\ E_3 \end{bmatrix}$$

$$= \begin{bmatrix} \epsilon & 0 & 0 \\ 0 & \epsilon & 0 \\ 0 & 0 & \epsilon \end{bmatrix} \begin{bmatrix} E_1 \\ E_2 \\ E_3 \end{bmatrix} = \epsilon \mathbf{E}, \tag{1.3-2}$$

where the subscripts 1, 2, and 3 refer to the x, y, and z components of the corresponding vector. In Eq. (1.3-2), I denotes the identity matrix and ϵ is a 3×3 diagonal matrix having nonzero elements each equal to ϵ.

In general, a 3×1 vector (such as $\mathbf{E}$) can be mapped onto another (such as $\mathbf{D}$) by a matrix transformation (such as ϵ). In fact, in an anisotropic medium, the ϵ matrix is no longer proportional to the identity matrix I and can have the form

$$\epsilon = \begin{bmatrix} \epsilon_{11} & \epsilon_{12} & \epsilon_{13} \\ \epsilon_{21} & \epsilon_{22} & \epsilon_{23} \\ \epsilon_{31} & \epsilon_{32} & \epsilon_{33} \end{bmatrix} = [\epsilon_{ij}], \qquad i = 1, 2, 3, j = 1, 2, 3. \quad (1.3\text{-}3)$$

Then, from Eqs. (1.3-2) and (1.3-3), we can write the elements of the vector $\mathbf{D}$ in terms of the elements of $\mathbf{E}$ as

$$\cdot \begin{bmatrix} D_1 \\ D_2 \\ D_3 \end{bmatrix} = \begin{bmatrix} \epsilon_{11} & \epsilon_{12} & \epsilon_{13} \\ \epsilon_{21} & \epsilon_{22} & \epsilon_{23} \\ \epsilon_{31} & \epsilon_{32} & \epsilon_{33} \end{bmatrix} \begin{bmatrix} E_1 \\ E_2 \\ E_3 \end{bmatrix} \qquad (1.3\text{-}4)$$

or, in short, as

$$D_i = \sum_{j=1}^{3} \epsilon_{ij} E_j, \qquad i = 1, 2, 3. \qquad (1.3\text{-}5)$$

Often, it is customary to drop the summation sign and write

$$D_i = \epsilon_{ij} E_j, \qquad (1.3\text{-}6)$$

where the summation is implied wherever there are repeated indices on the same side of an equation. This is called the *Einstein summation convention*.

Some of the familiar operations on vectors can be elegantly rewritten using the summation convention. For instance,

$$C = \mathbf{A} \cdot \mathbf{B} = A_i B_j \delta_{ij} = A_i B_i, \qquad (1.3\text{-}7)$$

$$F_i = (\mathbf{A} \times \mathbf{B})_i = e_{ijk} A_j B_k. \qquad (1.3\text{-}8)$$

In Eq. (1.3-7), δ_{ij} is the *Kronecker delta function*, defined as

$$\delta_{ij} = \begin{cases} 1 & \text{if } i = j, \\ 0 & \text{if } i \neq j. \end{cases} \qquad (1.3\text{-}9)$$

In Eq. (1.3-8), e_{ijk} is the *Levi–Civita tensor* and is defined as

$$e_{ijk} = \begin{cases} +1 & \text{if } (ijk) \text{ is an even permutation of } (123) \\ & [\text{e.g., } (123), (231), (312)], \\ -1 & \text{if } (ijk) \text{ is an odd permutation of } (123) \\ & [\text{e.g., } (132), (321), (213)], \\ 0 & \text{otherwise (some or all subscripts are equal).} \end{cases}$$

(1.3-10)

It can be shown that ϵ_{ij} is an *invariant bilinear function* of two dimensions. This means that we can find a constant T such that

$$T = \epsilon_{ij}\xi_i\eta_j, \qquad \epsilon_{ij} \text{ constant,} \qquad (1.3\text{-}11)$$

where ξ_i, η_j are various components of arbitrary unit vectors. Because we have two indices i and j characterizing ϵ_{ij}, it is of dimension, or *rank*, 2. A more technical name for ϵ_{ij} is "a *tensor* of rank 2." Extending the preceding discussion, we can define a Cartesian tensor $\epsilon_{ij\ldots n}$ of rank γ to be an invariant multilinear function of γ dimensions, that is, there exists a T such that

$$T = \underbrace{\epsilon_{ij\ldots n}}_{\gamma \text{ indices}} \underbrace{\xi_i\eta_j \cdots \lambda_n}_{\gamma \text{ factors}}, \qquad (1.3\text{-}12)$$

where $\xi_i, \eta_j, \ldots, \lambda_n$ are various components of arbitrary unit vectors. Note that a scalar is a tensor of tank 0, whereas a vector is a tensor of rank 1.

To step down from the level of abstraction, let us consider a simple illustration. Suppose that in Eq. (1.3-11), $\xi_i = \eta_i = E_i/|E_i|$, where the E_is are the components of an electric field. Then $T = \epsilon_{ij}E_iE_j/|E_i||E_j|$ and is reminiscent of the electromagnetic energy density if the ϵ_{ij}s denote permittivities. Now, the energy density is a coordinate-free quantity, which means that it remains invariant, irrespective of the coordinate system in which the E_is are specified. Hence, ϵ_{ij} is called the permittivity tensor. We will return to the permittivity (or dielectric) tensor in Chapter 6 when we discuss wave propagation in anisotropic media.

1.4 Comments on Nomenclature

In most of this text, we will be discussing wave propagation in one way or another. All wavefunctions, for example, components of the electric field, magnetic field, and so forth, will be generically denoted as $\psi(x, y, z, t)$, unless otherwise stated, and will be assumed to be real. In most cases of optical wave propagation, the temporal spectrum of ψ will be centered around a carrier frequency ω_0. Thus, we will write

$$\psi(x, y, z, t) = \text{Re}\left[\psi_p(x, y, z, t)e^{j\omega_0 t}\right], \qquad (1.4\text{-}1)$$

where $\psi_p(x, y, z, t)$ is called a *phasor* and is, in general, a complex quantity. In many optics texts, ψ_p is referred to as the *complex amplitude*. The complex phasor is often a "slowly varying" function of t, in the sense that

$$\frac{|\partial\psi_p/\partial t|}{|\psi_p|} \ll \omega_0. \qquad (1.4\text{-}2)$$

As stated in Section 1.1, propagating waves in the $+z$ direction have the form $\exp[j(\omega t - kz)]$. If the carrier frequency is ω_0, the propagation constant is $k_0 = \omega_0/c$ in vacuum. We will alternatively denote ψ by

$$\psi(x, y, z, t) = \text{Re}\left[\psi_e(x, y, z, t)\exp[j(\omega_0 t - k_0 z)]\right], \qquad (1.4\text{-}3)$$

where $\psi_e(x, y, z, t)$ is called the *complex envelope*. This is often a slowly varying function of t and z, the fast variations with respect to these variables having been incorporated into $\exp[j(\omega_0 t - k_0 z)]$. Comparing Eqs. (1.4-1) and (1.4-3), we note that the complex envelope $\psi_e(x, y, z, t)$ and the phasor $\psi_p(x, y, z, t)$ are related through the following equation:

$$\psi_p(x, y, z, t) = \psi_e(x, y, z, t)e^{-jk_0 z}. \qquad (1.4\text{-}4)$$

The representations of a real wave function ψ in terms of the phasor ψ_p and the complex envelope ψ_e, and the interrelationship between them, will be used repeatedly throughout the text.

Problems

1.1 Verify Fourier-transform pairs 4, 6, and 7 in Table 1.3.

1.2 From first principles, find the Fourier transform of the following:

(a) the signum function sgn(x, y), defined by

$$\text{sgn}(x, y) = \text{sgn}(x)\text{sgn}(y),$$

where

$$\text{sgn}(x) = \begin{cases} 1, & x > 0, \\ 0, & x = 0, \\ -1, & x < 0; \end{cases}$$

(b) sech(x/x_0)sech(y/y_0).

1.3 Using Tables 1.2 and 1.3, determine the Fourier transform of the signal $f(x, y) = f(x) \times 1$, where (a) $f(x)$ is shown in Figure P1.3(a) and (b) $f(x)$ is shown in Figure P1.3(b).

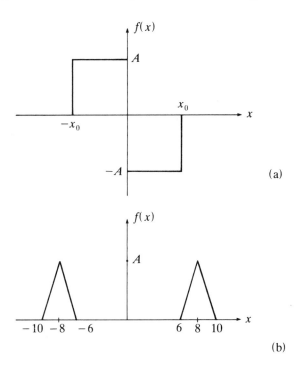

Figure P1.3

1.4 Verify that $f_i(x, y) * h(x, y) = h(x, y) * f_i(x, y)$ in Eq. (1.2-9).

1.5 If $G_i(k_x, k_y) = \mathscr{F}_{xy}\{g_i(x, y)\}$, where $i = 1, 2, 3$, determine the Fourier transforms of the following:
(a) $[g_1(x, y) * g_2(x, y)]g_3(x, y)$;
(b) $[g_1(x, y)g_2(x, y)] * g_3(x, y)$.

1.6 For a linear space-invariant optical system, the input $f_i(x, y)$ and output $f_o(x, y)$ signals are given by the integral relation (called the *Hilbert transform*),

$$f_o(x, y) = \frac{1}{\pi^2} \int_{-\infty}^{\infty} \int_{-\infty}^{\infty} \frac{f_i(\alpha, \beta)}{(x - \alpha)(y - \beta)} \, d\alpha \, d\beta.$$

Determine the transfer function $H(k_x, k_y)$ between the input and the output.

1.7 Prove *Parseval's theorem*:

$$\int_{-\infty}^{\infty} \int_{-\infty}^{\infty} |f(x, y)|^2 \, dx \, dy = \frac{1}{4\pi^2} \int_{-\infty}^{\infty} \int_{-\infty}^{\infty} |F(k_x, k_y)|^2 \, dk_x \, dk_y.$$

1.8 Using the Einstein convention and properties of the Kronecker delta and Levi–Civita tensors, prove the following vector identities:
(a) $\mathbf{A} \cdot (\mathbf{B} \times \mathbf{C}) = \mathbf{B} \cdot (\mathbf{C} \times \mathbf{A}) = \mathbf{C} \cdot (\mathbf{A} \times \mathbf{B})$;
(b) $\mathbf{A} \times (\mathbf{B} \times \mathbf{C}) = \mathbf{B}(\mathbf{A} \cdot \mathbf{C}) - \mathbf{C}(\mathbf{A} \cdot \mathbf{B})$.

1.9 Using the Einstein convention, express (a) ∇v (b) $\nabla \cdot \mathbf{A}$ (c) $\nabla \times \mathbf{A}$ (d) $\nabla^2 \mathbf{A}$ in a compact form. v denotes a scalar function.

References
1.1 Butkov, E. (1968). *Mathematical Physics*. Addison-Wesley, *Reading, Massachusetts*.
1.2 Korpel, A. and P. P. Banerjee (1984). *Proc. IEEE* **72** 1109.
1.3 Stremler, F. G. (1982). *Introduction to Communication Systems*. Addison-Wesley, Reading, Massachusetts.

Chapter 2 Geometrical Optics

In *geometrical* optics, we view light as particles of energy travelling through space. These particles follow trajectories that we call *rays*. We can describe an optical system comprising elements such as mirrors and lenses by tracing the rays through the system. In vacuum or free space, the speed of light particles is a constant, approximately given by $c = 3 \times 10^8$ m/s. The speed of light in a transparent linear, homogeneous, isotropic material, which we term v, is again a constant but less than c. This constant is a physical characteristic, or signature, of the material. The ratio c/v is called the *index of refraction n* of the material. For precise definitions of the words linear, homogeneous and isotropic, we refer readers to Chapter 3. In this chapter, we will restrict ourselves to linear isotropic media.

We will derive the laws of geometrical optics, namely, reflection and refraction, using a simple axiom known as *Fermat's principle*. This is an extremum principle from which we can trace the rays in a general optical medium. Based on the laws of reflection and refraction, we will introduce a matrix approach to analyze ray propagation through an optical system.

Geometrical optics is a special case of *wave* or *physical* optics, which will be our focus through the rest of the chapters in the book. Specifically, as we shall see in Chapter 3, we recover geometrical optics by taking the limit in which the wavelength of light approaches zero. In this limit, diffraction and the wave nature of light are absent.

2.1 Fermat's Principle

In classical mechanics, Hamilton's principle of least action provides a recipe to find the optimum displacement of a conservative system from one coordinate to another [Goldstein (1950)]. Similarly, in optics we have *Fermat's principle*, which states that *the path a ray of light follows is an extremum in comparison with the nearby paths*. In Section 2.2, we will use Fermat's principle to derive the laws of geometrical optics.

We now give a mathematical enunciation of Fermat's principle. Let $n(x, y, z)$ represent a position-dependent refractive index. Then,

$$\frac{ds}{c/n} = \frac{n\,ds}{c}$$

represents the time taken to traverse the geometric path ds in a medium of refractive index n. Thus, the time taken by the ray to traverse a path $\mathscr{C}$ between points A and B (see Figure 2.1) is

$$\frac{1}{c} \int_{\substack{A \to B \\ \mathscr{C}}} n(x, y, z)\,ds.$$

This integral is called the *optical path length* (OPL). According to Fermat's principle, the ray follows the path for which the OPL is an extremum, that is,

$$\delta(\text{OPL}) = \delta \int_{\substack{A \to B \\ \mathscr{C}}} n\,ds = 0. \qquad (2.1\text{-}1a)$$

The δ variation of the integration means that we find the partial differentials of the integral with respect to the free parameters in the integral. This will become clear in the next section, where we

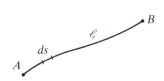

Figure 2.1 A ray of light traversing a path $\mathscr{C}$ between points A and B.

derive the laws of reflection and refraction when we have a common boundary between two media of different refractive indices. *In a homogenous medium* (i.e., in a medium with a constant refractive index), *the rays are straight lines.*

We can also restate Fermat's principle as a *principle of least time.* To see this, we divide Eq. (2.1-1a) by c to get

$$\frac{1}{c}\delta\int_{\underset{\ell}{A\to B}} n\,ds = 0. \qquad (2.1\text{-}1b)$$

We remark that Eq. (2.1-1b) is *incorrectly* called the least-time principle. To quote Feynman [Feynman, Leighton, and Sands (1963)], Eq. (2.1-1b) really means that "if we make a small change...in the ray in any manner whatever, say in the location at which it comes to the mirror, or the shape of the curve, or anything, there will be no first order change in the time; there will be only a second order change in the time."

2.2 Reflection and Refraction

When a ray of light is incident on the boundary MM' separating two different media, as in Figure 2.2, observation shows that part of the light is reflected back into the first medium, whereas the rest of the light is refracted as it enters the second medium. The directions taken by these rays are described by the laws of reflection and refraction. We will now use Fermat's principle to derive the two laws.

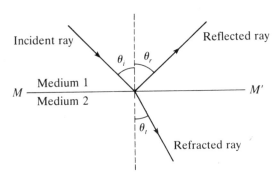

Figure 2.2 Reflected and refracted rays for light incident at the interface of two media.

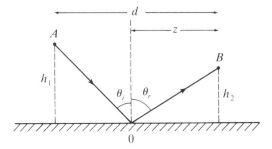

Figure 2.3 Incident (AO) and reflected (OB) rays.

Consider a reflecting surface as shown in Figure 2.3. Light from point A is deflected to point B by this surface, forming the angle of incidence θ_i and the angle of reflection θ_r, measured from the normal to the surface. The time required for the ray of light to travel the path $AO + OB$ is given by $t = (AO + OB)/v$, where v is the velocity of light in the medium containing the points AOB. The medium is considered isotropic for convenience. From the geometry, we find

$$t(z) = \frac{1}{v}\left[\{h_1^2 + (d - z)^2\}^{1/2} + \{h_2^2 + z^2\}^{1/2}\right]. \quad (2.2\text{-}1)$$

According to the least-time principle, light will find a path that extremizes $t(z)$ with respect to variations in z. We thus set $dt(z)/dz = 0$ to get

$$\frac{d - z}{\{h_1^2 + (d - z)^2\}^{1/2}} = \frac{z}{\{h_2^2 + z^2\}^{1/2}}, \quad (2.2\text{-}2)$$

or

$$\sin\theta_i = \sin\theta_r, \quad (2.2\text{-}3a)$$

so that

$$\theta_i = \theta_r. \quad (2.2\text{-}3b)$$

We can readily check that the second derivative of $t(z)$ is positive

Geometrical Optics

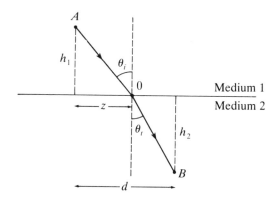

Figure 2.4 Incident (AO) and refracted (OB) rays.

so that the result obtained indeed corresponds to the *least*-time principle. Equation (2.2-3b) states that the angle of incidence is equal to the angle of reflection. In addition, Fermat's principle also demands that the incident ray, the reflected ray, and the normal all be in the same plane, called the *plane of incidence*.

Let us now use Fermat's principle to analyze refraction as illustrated in Figure 2.4. θ_i and θ_t are the angles of incidence and transmission, respectively, measured once again from the normal to the interface. The time taken by the light to travel the distance AOB is

$$t(z) = \frac{AO}{v_1} + \frac{OB}{v_2} = \frac{\{h_1^2 + z^2\}^{1/2}}{v_1} + \frac{\{h_2^2 + (d-z)^2\}^{1/2}}{v_2}, \quad (2.2\text{-}4)$$

where v_1 and v_2 are the light velocities in media 1 and 2, respectively. In order to minimize $t(z)$, we set

$$\frac{dt}{dz} = \frac{z}{v_1\{h_1^2 + z^2\}^{1/2}} - \frac{d-z}{v_2\{h_2^2 + (d-z)^2\}^{1/2}} = 0. \quad (2.2\text{-}5)$$

Using the geometry of the problem, we conclude that

$$\frac{\sin \theta_i}{v_1} = \frac{\sin \theta_t}{v_2}. \quad (2.2\text{-}6a)$$

Now $v_1 = c/n_1$ and $v_2 = c/n_2$, where n_1 and n_2 are the refractive indices of media 1 and 2, respectively. Equation (2.2-6a) can thus be restated as

$$\frac{\sin \theta_i}{\sin \theta_t} = \frac{n_2}{n_1}, \qquad (2.2\text{-}6b)$$

where n_2/n_1 is the *relative refractive index* of medium 2 with respect to medium 1. Equation (2.2-6b) is called *Snell's law of refraction*. Again, as in reflection, the incident ray, the refracted ray, and the normal all lie in the same plane of incidence. Snell's law shows that when a light ray passes obliquely from a medium of smaller refractive index into one that has a larger refractive index, it is bent toward the normal. Conversely, if the ray of light travels into a medium with a lower refractive index, it is bent away from the normal. For the latter case, it is possible to visualize a situation where the refracted ray is bent away from the normal by exactly 90°. Under this situation, the angle of incidence is called the *critical angle* θ_c, and is given by

$$\sin \theta_c = \frac{n_2}{n_1}. \qquad (2.2\text{-}7)$$

When the incident angle is greater than the critical angle, the ray originating in medium 1 is totally reflected back into medium 1. This phenomenon is called *total internal reflection* (TIR). The optical fiber uses this principle of total reflection to guide light, and the mirage on a hot summer day is a phenomenon due to the same principle.

We present the electromagnetic treatment of reflection and refraction, including total internal reflection, in the next chapter, where we derive the Fresnel equations.

2.3 Refraction in an Inhomogeneous Medium

In the previous section, we discussed refraction between two media with different refractive indices, that is, possessing a discrete inhomogeneity. Consider now a medium comprising a continuous set of thin slices of media of different refractive indices, as shown in

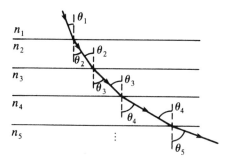

Figure 2.5 Rays in a layered medium in which the refractive index is piece-wise continuous.

Figure 2.5. At every interface, the light ray satisfies Snell's law according to

$$n_1 \sin \theta_1 = n_2 \sin \theta_2 = n_3 \sin \theta_3 = \cdots. \qquad (2.3\text{-}1)$$

Thus, we may put

$$n \sin \theta = \text{constant} = n_1 \sin \theta_1, \qquad (2.3\text{-}2)$$

where $n(x)$ and $\theta(x)$ stand for the refractive index and the angle in a general layer, respectively, at location x. In the limiting case of a continuous variation of the refractive index, with position (namely x), which defines an *inhomogeneous medium*, the piecewise linear trajectory of the ray becomes a continuous curve, as shown in Figure 2.6. If ds represents the infinitesimal arc length along the curve, then

$$(ds)^2 = (dx)^2 + (dz)^2, \qquad (2.3\text{-}3)$$

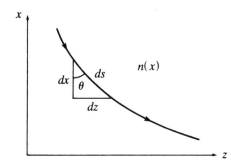

Figure 2.6 The path of a ray in a medium with a continuous inhomogeneity.

where we restrict ourselves to two dimensions. Also, from Figure 2.6,

$$\frac{dz}{ds} = \sin \theta. \tag{2.3-4}$$

Combining Eqs. (2.3-2), (2.3-3), and (2.3-4), we obtain

$$\left(\frac{dx}{dz}\right)^2 = \frac{n^2(x)}{n_1^2 \sin^2 \theta_1} - 1, \tag{2.3-5a}$$

or, alternatively, by differentiating with respect to z,

$$\frac{d^2x}{dz^2} = \frac{1}{2n_1^2 \sin^2 \theta_1} \frac{dn^2(x)}{dx}. \tag{2.3-5b}$$

The solution of Eq. (2.3-5a) [or Eq. (2.3-5b)] gives the direction a ray will take in a medium with a refractive index variation $n(x)$ and is a special form of the ray, or *eikonal*, equation. (More on the eikonal equation follows in the next chapter.)

Example 2.1 Homogenous Medium

$$n(x) = \text{constant}.$$

From Eq. (2.3-5a), we observe that the solution is that of a straight line, as expected.

Example 2.2 Square-Law Medium

$$n^2(x) = n_0^2 - \tilde{n}x^2. \tag{2.3-6}$$

Substituting for n^2 from Eq. (2.3-6) into Eq. (2.3-5b) yields

$$\frac{d^2x}{dz^2} = -\frac{\tilde{n}}{n_0^2 \sin^2 \theta_1}x(z), \tag{2.3-7}$$

where θ_1 is the angle the ray makes with the x (transverse) axis at $x = 0$. The solution of Eq. (2.3-7) is of the form

$$x(z) = A \sin\left\{ \frac{\tilde{n}^{1/2}}{n_0 \sin \theta_1} z + \phi_0 \right\}, \qquad (2.3\text{-}8)$$

where the constants A and ϕ_0 can be determined from the initial position and slope of the ray. Note that rays with smaller launching angles α $(= \pi/2 - \theta_1)$ have a larger period; however, in the paraxial approximation (i.e., for small launching angles), all the ray paths have approximately the same period.

The case discussed in Example 2.2 approximately explains the mechanism of light propagation through *graded-index optical fibers*. Wave propagation through such fibers will be discussed in Chapter 3.

2.4 Matrix Methods in Paraxial Optics

In this section, we consider how matrices may be used to describe ray propagation through optical systems comprising, for instance, a succession of spherical refracting and/or reflecting surfaces all centered on the same axis, which is called the *optical axis*. Unless otherwise stated, we will take the optical axis to be along the z axis. As we will see shortly, the "coordinates" of a ray at a certain plane perpendicular to the optical axis can be specified by a vector that contains the information of the position and direction of the ray. It would therefore be convenient if, given this information, we can find the coordinates of the ray at any other plane, again normal to the optical axis, by means of successive operators acting on the initial ray coordinate vector, with each operator being characteristic of the optical element through which the ray travels. We can represent these operators by matrices. The advantage of this matrix formalism is that any ray, during its propagation through the optical system, can be tracked by successive matrix multiplications, which can be easily programmed on a digital computer. This representation of geometrical optics is elegant and powerful and is widely used in optical element designs.

We will only consider *paraxial* rays, implying rays that are close to the optical axis and whose angular deviation from it is small enough that the sine and tangent of the angles may be approximated by the angles themselves. This paraxial approximation guarantees that all paraxial rays starting from a given *object point* intersect at another point (the *image point*) after passage through the optical system. Nonparaxial rays may not give rise to a single image point; this phenomenon, called *aberration*, is outside the scope of this book. The optics of paraxial imaging is also sometimes called *Gaussian optics*.

In what follows, we will first develop the matrix formalism for paraxial ray propagation, or *ray transfer*, and examine some of the properties of ray transfer matrices. We then consider several illustrative examples. For instance, we examine the imaging properties of lenses and mirrors and derive the rules for ray tracing through an optical system. Finally, we use the matrix method to analyze ray propagation within an optical resonator which forms an integral part of a laser system. This is considered in detail in Chapter 7.

2.4.1 The Ray Transfer Matrix

Consider the propagation of a paraxial ray through an optical system as shown in Figure 2.7. Restricting ourselves to one transverse direction (x), a ray at a given cross section or plane may be specified by its height x from the optical axis and by its angle or slope θ that it makes with the axis. Thus, the quantities (x, θ) represent the coordinates of the ray for a given z-constant plane. However, instead of specifying the angle the ray makes with the z

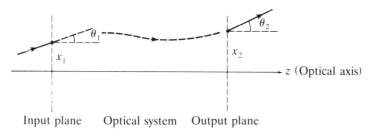

Figure 2.7 Reference planes of an optical system.

axis, it is customary to replace the corresponding θ by $\varpi = n\theta$, where n is the refractive index at the z-constant plane.

In Figure 2.7, the ray passes through the input plane with coordinates $(x_1, \varpi_1 = n_1\theta_1)$, then through the optical system, and finally through the output plane with coordinates $(x_2, \varpi_2 = n_2\theta_2)$. In the paraxial approximation, the corresponding output quantities are linearly dependent on the input quantities. We can, therefore, represent the transformation from the input to the output in matrix form as

$$\begin{pmatrix} x_2 \\ \varpi_2 \end{pmatrix} = \begin{pmatrix} \mathscr{A} & \mathscr{B} \\ \mathscr{C} & \mathscr{D} \end{pmatrix} \begin{pmatrix} x_1 \\ \varpi_1 \end{pmatrix}. \tag{2.4-1}$$

The $\mathscr{ABCD}$ matrix in Eq. (2.4-1) is called the *ray transfer matrix* and, as we shall see later, it can be made up of many matrices to account for the effects of a ray passing through various optical elements. We can consider these matrices as operators successively acting on the input ray coordinate vector. We state here that the determinant of the ray transfer matrix equals unity, that is, $\mathscr{AD} - \mathscr{BC} = 1$. This will become clear after we derive the translation, refraction, and reflection matrices.

Let us now investigate the general properties of an optical system from the $\mathscr{ABCD}$ matrix.

Property 1: If $\mathscr{D} = 0$, we have from Eq. (2.4-1) that $\varpi_2 = \mathscr{C}x_1$. This means that all rays crossing the input plane at the same point, namely, x_1, emerge at the output plane making the same angle with the axis, no matter at what angle they enter the system. The input plane is called the *front focal plane* of the optical system [see Figure 2.8(a)].

Property 2: If $\mathscr{B} = 0$, $x_2 = \mathscr{A}x_1$ [from Eq. (2.4-1)]. This means that all rays passing through the input plane at the same point (x_1) will pass through the same point (x_2) in the output plane [see Figure 2.8(b)]. The input and output planes are called the *object* and *image planes*, respectively. In addition, $\mathscr{A} = x_2/x_1$ gives the *magnification* produced by the system.

Furthermore, by inverting the $\mathscr{ABCD}$ matrix and from the fact that $\mathscr{AD} - \mathscr{BC} = 1$, we note from Eq. (2.4-1) that $x_1 = \mathscr{D}x_2 = (1/\mathscr{A})x_2$, because $\mathscr{B} = 0$. The implication of this is

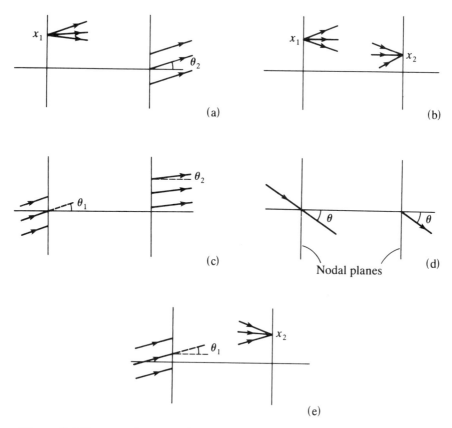

Figure 2.8 Rays at input and output planes for (a) $\mathscr{D} = 0$, (b) $\mathscr{B} = 0$, (c) $\mathscr{C} = 0$, (d) the case when the planes are nodal planes, and (e) $\mathscr{A} = 0$.

that the point x_2 is imaged at x_1 with magnification $1/\mathscr{A}$. Hence, the two planes containing x_1 and x_2 are called *conjugate planes*. Moreover, if $\mathscr{A} = 1$, that is, the magnification between the two conjugate planes is unity, these two planes are called the *unit*, or *principal*, *planes*. The points of intersection of the unit planes with the optical axis are the *unit*, or *principal*, *points*. The principal points constitute one set of *cardinal points*.

Property 3: If $\mathscr{C} = 0$, $v_2 = \mathscr{D} v_1$. This means that all the rays entering the system parallel to one another will also emerge parallel, albeit in a new direction [see Figure 2.8(c)]. In addition, $\mathcal{D}(n_1/n_2) = \theta_2/\theta_1$ gives the *angular magnification* produced by the system.

If $\mathcal{D} = n_2/n_1$, we have unity angular magnification, that is, $\theta_2/\theta_1 = 1$. In this case, the input and output planes are referred to as the *nodal planes*. The intersections of the nodal planes with the optical axis are called the *nodal points* [see Figure 2.8(d)]. The nodal points constitute the other set of cardinal points.

Property 4: If $\mathcal{A} = 0$, $x_2 = \mathcal{B}v_1$. This means that all rays entering the system at the same angle will pass through the same point at the output plane. The output plane is the *back focal plane* of the system [see Figure 2.8(e)].

2.4.2 Translation and Refraction Matrices

When a ray passes through an optical system, there are usually two types of processes, translation and refraction (and, sometimes, reflection; this is treated later), that we need to consider in order to determine the progress of the ray. As the rays propagate through a homogeneous medium, they undergo a translation process. In order to specify the translation, we need to know the thickness of the medium and its refractive index. However, when a ray strikes an interface between two regions of different refractive indices, it undergoes refraction. To determine how much bending the ray undergoes, we need to know the radius of curvature of the boundary and the values of the refractive indices of the two regions. We shall investigate the effect each of these two processes has on the coordinates of a ray between the input and the output planes. In fact, we will derive the ray transfer matrices for the two processes.

Figure 2.9 shows a ray travelling a distance d in a homogeneous medium of refractive index n. Because the medium is homogeneous, the ray travels in a straight line. The set of equations of translation by a distance d is

$$x_2 = x_1 + d \tan \theta_1 \approx x_1 + \theta_1 d, \qquad (2.4\text{-}2a)$$

$$n\theta_2 = n\theta_1 \quad \text{or} \quad v_2 = v_1. \qquad (2.4\text{-}2b)$$

These equations relate the output coordinates of the ray with its

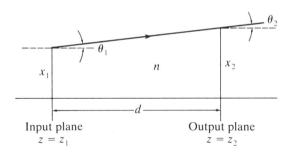

Figure 2.9 A ray in a homogenous medium of refractive index n_o.

input coordinates. We can express this transformation in matrix form as

$$\begin{pmatrix} x_2 \\ \theta_2 \end{pmatrix} = \begin{pmatrix} 1 & d/n \\ 0 & 1 \end{pmatrix} \begin{pmatrix} x_1 \\ \theta_1 \end{pmatrix}. \tag{2.4-3}$$

The 2×2 ray transfer matrix, called the *translation matrix* $\mathscr{T}$, is defined as

$$\mathscr{T} = \begin{pmatrix} 1 & d/n \\ 0 & 1 \end{pmatrix}. \tag{2.4-4}$$

Note that its determinant is unity.

We now adopt the following convention: When light rays travel a distance d from the plane $z = z_1$ to the plane $z = z_2$ (see Figure 2.9), $z_2 - z_1$ will be taken to be positive for a ray travelling in the $+z$ direction and negative for a ray travelling in the $-z$ direction. Therefore, in the latter case, we take the refractive index of the medium to be negative so that the value of $(z_2 - z_1)/n$ in the translation matrix will remain positive.

We next study the effect of a spherical surface separating two regions of refractive indices n_1 and n_2 as shown in Figure 2.10. The center of the curved surface is at C and its radius of curvature is R. The ray strikes the surface at the point A and gets refracted. Note that the radius of curvature of the surface will be taken as positive (negative) if the center C of curvature lies to the right (left) of the surface. Let x be the distance from A to the axis. Then the

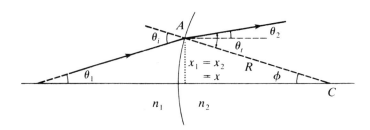

Figure 2.10 Ray trajectory during refraction at a spherical surface.

angle ϕ subtended at the center C becomes

$$\sin \phi = \frac{x}{R} \simeq \phi \quad \text{(paraxial approximation).} \quad (2.4\text{-}5)$$

We see that, in this case, the height of the ray at A before and after refraction is the same, that is, $x_2 = x_1$. We therefore need to obtain the relationship for $\varkappa_2$ in terms of x_1 and $\varkappa_1$. Applying Snell's law [Eq. (2.2-6b)] and using the paraxial approximation, we have

$$n_1 \theta_i = n_2 \theta_t. \quad (2.4\text{-}6)$$

From geometry, we know from Figure 2.10 that $\theta_i = \theta_1 + \phi$ and $\theta_t = \theta_2 + \phi$. Hence,

$$n_1 \theta_i = \varkappa_1 + \frac{n_1 x_1}{R}, \quad (2.4\text{-}7a)$$

$$n_2 \theta_t = \varkappa_2 + \frac{n_2 x_2}{R}. \quad (2.4\text{-}7b)$$

Using Eqs. (2.4-6), (2.4-7a), and (2.4-7b) and the fact that $x_1 = x_2$, we obtain

$$\varkappa_2 = \frac{n_1 - n_2}{R} x_1 + \varkappa_1. \quad (2.4\text{-}8)$$

The matrix–vector equation relating the coordinates of the ray after refraction to those before refraction becomes

$$\begin{pmatrix} x_2 \\ v_2 \end{pmatrix} = \begin{pmatrix} 1 & 0 \\ -p & 1 \end{pmatrix} \begin{pmatrix} x_1 \\ v_1 \end{pmatrix}, \qquad (2.4\text{-}9a)$$

where the quantity p, given as

$$p = \frac{n_2 - n_1}{R}, \qquad (2.4\text{-}9b)$$

is termed the *refracting power* of the spherical surface. With R measured in meters, the unit of p is called the *diopter*. If an incident ray is made to converge (diverge) by a surface, the power will be assumed to be positive (negative) in sign. The 2×2 transfer matrix is called the *refraction matrix* $\mathscr{R}$ and it describes refraction at A for the spherical surface:

$$\mathscr{R} = \begin{pmatrix} 1 & 0 \\ -p & 1 \end{pmatrix}. \qquad (2.4\text{-}10)$$

Note that the determinant of $\mathscr{R}$ is also unity.

2.4.3 Illustrative Examples

Example 2.3 Plane-Parallel Layers
Consider a medium of thickness d, divided into two regions of thicknesses d_1 and d_2, each having the same refractive index n, as shown in Figure 2.11. The system of equations relating the

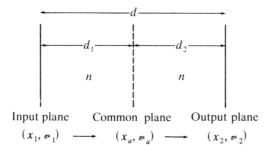

Figure 2.11 Plane parallel layers of thicknesses d_1 and d_2.

Geometrical Optics

output coordinates to the input coordinates is given by

$$\begin{pmatrix} x_2 \\ v_2 \end{pmatrix} = \begin{pmatrix} 1 & \dfrac{d_2}{n} \\ 0 & 1 \end{pmatrix} \begin{pmatrix} x_a \\ v_a \end{pmatrix}$$

$$= \begin{pmatrix} 1 & \dfrac{d_2}{n} \\ 0 & 1 \end{pmatrix} \begin{pmatrix} 1 & \dfrac{d_1}{n} \\ 0 & 1 \end{pmatrix} \begin{pmatrix} x_1 \\ v_1 \end{pmatrix}. \qquad (2.4\text{-}11)$$

The overall system transfer matrix $\mathscr{T}$ can be written as

$$\mathscr{T} = \mathscr{T}_2 \mathscr{T}_1 = \begin{pmatrix} 1 & \dfrac{d_2}{n} \\ 0 & 1 \end{pmatrix} \begin{pmatrix} 1 & \dfrac{d_1}{n} \\ 0 & 1 \end{pmatrix}$$

$$= \begin{pmatrix} 1 & \dfrac{d_1 + d_2}{n} \\ 0 & 1 \end{pmatrix}, \qquad (2.4\text{-}12)$$

as expected. Note that the overall system matrix $\mathscr{T}$ is expressed in terms of the product of the two individual matrices $\mathscr{T}_1$ and $\mathscr{T}_2$ written in order from right to left. The order of the matrix multiplication is important, as matrix multiplication is not commutative in general.

A similar situation applies when a region of thickness d comprises i layers, each having a thickness d_j and refractive index n_j. We can write the overall system matrix $\mathscr{T}$ as

$$\mathscr{T} = \mathscr{T}_i \cdots \mathscr{T}_2 \mathscr{T}_1 = \prod_{j=1}^{i} \begin{pmatrix} 1 & \dfrac{d_j}{n_j} \\ 0 & 1 \end{pmatrix} = \begin{pmatrix} 1 & \displaystyle\sum_{j=1}^{i} \dfrac{d_j}{n_j} \\ 0 & 1 \end{pmatrix}. \qquad (2.4\text{-}13)$$

In writing Eq. (2.4-13), we have neglected the refraction matrices at each plane of separation between successive layers, because the refraction matrix reduces to the identity matrix in the paraxial regime.

Example 2.4 A Single Lens

Consider a single lens as shown in Figure 2.12. It is evident that the system matrix for the lens consists of two refraction matrices and a translation matrix:

$$
\mathscr{S} = \begin{pmatrix} 1 & 0 \\ \dfrac{n_2 - n_1}{R_2} & 1 \end{pmatrix} \begin{pmatrix} 1 & \dfrac{d}{n_2} \\ 0 & 1 \end{pmatrix} \begin{pmatrix} 1 & 0 \\ \dfrac{n_1 - n_2}{R_1} & 1 \end{pmatrix}. \qquad (2.4\text{-}14)
$$

refraction at   translation   refraction at
surface 2   surface 1

For a *thin* lens in air, $d \rightarrow 0$ and $n_1 = 1$. Writing $n_2 = n$ for notational convenience, Eq. (2.4-14) becomes

$$
\mathscr{S} = \begin{pmatrix} 1 & 0 \\ -p_2 & 1 \end{pmatrix} \begin{pmatrix} 1 & 0 \\ 0 & 1 \end{pmatrix} \begin{pmatrix} 1 & 0 \\ -p_1 & 1 \end{pmatrix}, \qquad (2.4\text{-}15)
$$

where $p_1 = (n-1)/R_1$ and $p_2 = (1-n)/R_2$ are the refractive powers of surfaces 1 and 2, respectively. Note that the translation matrix degenerates into a unit matrix. Equation (2.4-15) can be rewritten as

$$
\mathscr{S} = \begin{pmatrix} 1 & 0 \\ -p_2 & 1 \end{pmatrix} \begin{pmatrix} 1 & 0 \\ -p_1 & 1 \end{pmatrix} = \begin{pmatrix} 1 & 0 \\ -(p_1 + p_2) & 1 \end{pmatrix} = \begin{pmatrix} 1 & 0 \\ -\dfrac{1}{f} & 1 \end{pmatrix},
$$

$$
(2.4\text{-}16)
$$

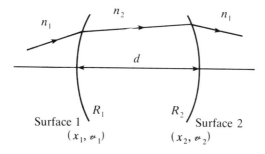

Figure 2.12 A single lens. The radii of curvature of the front and back surfaces are R_1 and R_2.

Geometrical Optics

where f is the *focal length* and is given explicitly by

$$\frac{1}{f} = (n - 1)\left(\frac{1}{R_1} - \frac{1}{R_2}\right). \qquad (2.4\text{-}17)$$

We will clarify the implication of the focal length in the following example, where we discuss ray tracing through a thin lens.

For $R_1 > 0$ and $R_2 < 0$ ($R_1 < 0$ and $R_2 > 0$), $f > 0$ ($f < 0$). If a ray of light is incident on the left surface of the lens parallel to the axis, the angle at which it emerges on the right surface may be found by using Eq. (2.4-1), with $v_1 = 0$, and the ray transfer matrix for the thin lens, as in Eq. (2.4-16). It follows that for $f > 0$ ($f < 0$), the ray bends toward (away from) the axis upon refraction through the lens. In the first case, the lens is called a *converging* (i.e., *convex*) lens, whereas in the second case, we have a *diverging* (i.e., *concave*) lens.

Example 2.5 Ray Tracing through a Single Thin Lens

In the previous example, we mentioned how rays parallel to the axis of a thin lens in air were bent toward or away from the axis after passing through a thin lens, depending upon whether the lens was converging or diverging. To carry this idea on a little further, consider the following cases:

(a) Ray travelling parallel to the axis

The input ray vector is $(x_1, 0)^T$, hence the output ray vector is given, using Eqs. (2.4-1) and (2.4-16), as $(x_1, -x_1/f)^T$. This ray now travels in a straight line at an angle $-x_1/f$ with the axis. This means that if x_1 is positive (or negative), the ray after refraction through the lens intersects the optical axis at a point a distance f behind the lens if the lens is converging ($f > 0$). This justifies calling f the focal length of the lens. All rays parallel to the optical axis in front of the lens converge behind the lens to a point called the *back focus* [see Figure 2.13(a)]. In the case of a diverging lens ($f < 0$), the ray after refraction diverges away from the axis as if it

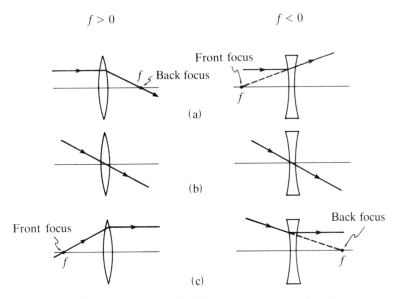

Figure 2.13 Ray tracing through thin converging and diverging lenses.

were coming from a point on the axis a distance $|f|$ in front of the lens. This point is called the *front focus*. This is also shown in Figure 2.13(a).

(b) Ray travelling through the center of the lens

The input ray vector is $(0, \varkappa_1)^T$, hence the output ray vector is given, using Eqs. (2.4-1) and (2.4-16), as $(0, \varkappa_1)^T$. This means that a ray travelling through the center of the lens will pass undeviated, as shown in Figure 2.13(b).

(c) Ray passing through the front focus of a converging lens

The input ray vector is given by $(x_1, x_1/f)^T$, so that the output ray vector is $(x_1, 0)^T$. This means that the output ray will be parallel to the axis, as shown in Figure 2.13(c).

In a similar way, we can show that for an input ray appearing to travel toward the back focus of a diverging lens, the output ray will be parallel to the axis.

Example 2.6 Imaging by a Single Thin Lens

Consider an object OO' located a distance d_o in front of a thin lens of focal length f, as shown in Figure 2.14. Assume that $(x_o, \varkappa_o)^T$ represents the coordinates of a ray originating from point O', and travelling toward the lens. Then the output ray coordinates $(x, \varkappa)$ at a distance z behind the lens can be written in terms of the input ray coordinates, two translation matrices, and the transfer matrix for the thin lens as

$$\begin{pmatrix} x \\ \varkappa \end{pmatrix} = \begin{pmatrix} 1 & z \\ 0 & 1 \end{pmatrix} \begin{pmatrix} 1 & 0 \\ -\dfrac{1}{f} & 1 \end{pmatrix} \begin{pmatrix} 1 & d_o \\ 0 & 1 \end{pmatrix} \begin{pmatrix} x_o \\ \varkappa_o \end{pmatrix} \qquad (2.4\text{-}18a)$$

$$= \begin{pmatrix} 1 - \dfrac{z}{f} & d_o + z - \dfrac{d_o z}{f} \\ -\dfrac{1}{f} & 1 - \dfrac{d_o}{f} \end{pmatrix} \begin{pmatrix} x_o \\ \varkappa_o \end{pmatrix}. \qquad (2.4\text{-}18b)$$

Assume for a moment that the object is a point source on the axis, that is, $x_o = 0$. Consider two rays emanating from the point object, one along the axis, the other at an angle $\varkappa_o$. Upon refraction through the lens, the on-axis ray will emerge undeviated, whereas the other ray will emerge at an angle $\varkappa$. We define the *image* of the point object to be the point where these two rays meet. Clearly, in this case, the image will be a point on the axis. We can calculate the distance $z = d_i$ along the axis behind the lens where the image will form by setting x_o and x equal to zero in Eq. (2.4-18b). This yields the celebrated thin-lens formula,

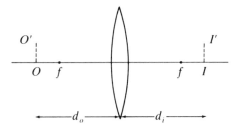

Figure 2.14 Imaging by a single lens.

form by setting x_o and x equal to zero in Eq. (2.4-18b). This yields the celebrated thin-lens formula,

$$\frac{1}{d_o} + \frac{1}{d_i} = \frac{1}{f}. \tag{2.4-19}$$

The *sign convention* for d_o and d_i is as follows: d_o is positive if the object is to the left of the lens. Now, if d_i is positive (negative), the image is to the right (left) of the lens and it is real (virtual).

Now, returning to Eq. (2.4-18b), we have (corresponding to the image plane) the relation

$$\begin{pmatrix} x_i \\ v_i \end{pmatrix} = \begin{pmatrix} 1 - \dfrac{d_i}{f} & 0 \\ -\dfrac{1}{f} & 1 - \dfrac{d_o}{f} \end{pmatrix} \begin{pmatrix} x_o \\ v_o \end{pmatrix}, \tag{2.4-20}$$

where we have written (x_i, v_i) for the coordinates of the ray at the image plane $z = d_i$. For $x_o \neq 0$, we obtain

$$\frac{x_i}{x_o} \triangleq M = 1 - \frac{d_i}{f} = \frac{f - d_i}{f} \tag{2.4-21a}$$

$$= \frac{f}{f - d_o} = -\frac{d_i}{d_o}, \quad \text{using Eq. (2.4-19),} \tag{2.4-21b}$$

where M is called the *magnification* of the system. If $M > 0 \, (< 0)$, the image is erect (inverted).

Example 2.7 Two-lens Combination

Consider two thin lenses of focal lengths f_1 and f_2, separated by a distance $d \, (< f_1 + f_2)$ in air (see Figure 2.15). We can write the overall transfer matrix for the two-lens system as

$$\mathcal{S} = \begin{pmatrix} 1 & 0 \\ -\dfrac{1}{f_2} & 1 \end{pmatrix} \begin{pmatrix} 1 & d \\ 0 & 1 \end{pmatrix} \begin{pmatrix} 1 & 0 \\ -\dfrac{1}{f_1} & 1 \end{pmatrix}. \tag{2.4-22}$$

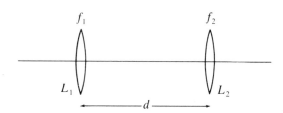

Figure 2.15 A two-lens system.

It is left as an exercise for the reader to show that the effective focal length f of an equivalent lens, located at a distance

$$d_2 = \frac{fd}{f_1} \qquad (2.4\text{-}23a)$$

in front of the position of the lens of focal length f_2, can be written as

$$\frac{1}{f} = \frac{1}{f_1} + \frac{1}{f_2} - \frac{d}{f_1 f_2}, \qquad (2.4\text{-}23b)$$

if the rays are travelling from the left to the right.

As another example of imaging using a two-lens system, let $f_1 = f_2 = f$ and consider an object $2f$ units to the left of lens L_1. Assume also $d = 4f$. Because the equivalent lens method is not applicable (because $d > f_1 + f_2$), we will solve the problem from first principles. The matrix chain from the object plane to the image plane becomes

$$\tilde{\mathscr{S}} = \begin{pmatrix} 1 & d_i \\ 0 & 1 \end{pmatrix} \mathscr{S} \begin{pmatrix} 1 & 2f \\ 0 & 1 \end{pmatrix}, \qquad (2.4\text{-}24a)$$

where $\mathscr{S}$ is defined in (2.4-22) and d_i (> 0) is the distance of the image plane to the right of lens L_2. From Eq. (2.4-24a),

$$\tilde{\mathscr{S}} = \begin{pmatrix} -3 + \dfrac{2d_i}{f} & -2f + d_i \\ \dfrac{2}{f} & 1 \end{pmatrix}, \qquad (2.4\text{-}24b)$$

from which we can find the position of the image by setting $B = -2f + d_i = 0$. Hence, $d_i = 2f$ and the magnification (A) is then equal to 1, implying a real erect image of the same size as the object. The result can be readily verified by drawing a simple ray diagram.

Example 2.8 Imaging by a Thick Lens; Cardinal Points

In Example 2.4, we wrote down the system matrix $\mathscr{S}$ for a thick lens in terms of the radii of curvature R_1 and R_2 of both surfaces and the thickness of the lens. In the discussion that follows, we will symbolically write the elements of $\mathscr{S}$ in Eq. (2.4-14) as

$$\mathscr{S} = \begin{pmatrix} a & b \\ c & d \end{pmatrix}. \tag{2.4-25}$$

As in the thin-lens case, we can study the imaging properties of a thick lens in air by considering the input and output planes to be located at distances d_1 and d_2 from the front and back refracting surfaces, respectively, as shown in Figure 2.16. We can find the

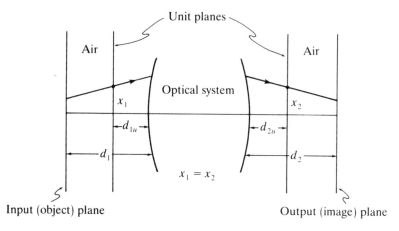

Figure 2.16 Unit planes for an optical system. A ray starting from any height x_1 from the input unit plane will cross the output unit plane at $x_2 = x_1$.

Geometrical Optics

output ray coordinates $(x_2, v_2)^T$ in terms of the input ray coordinates $(x_1, v_1)^T$ as

$$\begin{pmatrix} x_2 \\ v_2 \end{pmatrix} = \begin{pmatrix} 1 & d_2 \\ 0 & 1 \end{pmatrix} \begin{pmatrix} a & b \\ c & d \end{pmatrix} \begin{pmatrix} 1 & d_1 \\ 0 & 1 \end{pmatrix} \begin{pmatrix} x_1 \\ v_1 \end{pmatrix} \qquad (2.4\text{-}26a)$$

$$\underbrace{\qquad}_{\text{translation}} \quad \underbrace{\qquad}_{\text{thick lens}} \quad \underbrace{\qquad}_{\text{translation}}$$

$$= \begin{pmatrix} a + c\,d_2 & a\,d_1 + b + c\,d_1 d_2 + d\,d_2 \\ c & c\,d_1 + d \end{pmatrix} \begin{pmatrix} x_1 \\ v_1 \end{pmatrix}. \qquad (2.4\text{-}26b)$$

Once again, to find the location of the image (d_2), we consider a point object located on the axis in the input plane, that is, take $x_1 = 0$. Because the image should also be a point on the axis, we can set $x_2 = 0$ in Eq. (2.4-26b). This gives

$$a\,d_1 + b + c\,d_1 d_2 + d\,d_2 = 0, \qquad (2.4\text{-}27)$$

from which we can easily solve for d_2 in terms of d_1. Also, the image magnification can be found using Eqs. (2.4-26) and (2.4-27) as

$$\frac{x_2}{x_1} = (a + c\,d_2). \qquad (2.4\text{-}28)$$

It is instructive, at this point, to determine the cardinal points of our thick-lens imaging system. By setting $x_2/x_1 = 1$ and making use of the fact that the determinant of the matrix in Eq. (2.4-26b) is equal to unity, we find the locations of the two unit planes d_{1u} and d_{2u} as

$$d_{1u} = \frac{1 - d}{c}, \qquad d_{2u} = \frac{1 - a}{c}. \qquad (2.4\text{-}29)$$

The two unit planes are shown in Figure 2.16. Next, to find the nodal points, we set $v_1 = v_2$. From Eq. (2.4-26b), we find the locations of the two nodal points d_{1n} and d_{2n} as

$$d_{1n} = \frac{1 - d}{c} = d_{1u}, \qquad d_{2n} = \frac{1 - a}{c} = d_{2u}. \qquad (2.4\text{-}30)$$

Note that the nodal planes coincide with the unit planes when the media on both sides of the optical system have the same refractive index.

2.4.4 Extension of the Ray-Transfer Method to Reflecting Systems

According to our convention stated in Section 2.4.1, when light rays travel between planes $z = z_1$ and $z = z_2 > z_1$, $z_2 - z_1$ will be taken to be positive (negative) for a ray travelling in the $+z$ ($-z$) direction. Therefore, in the latter case, the refractive index of the medium must be taken as negative so that the value of $(z_2 - z_1)/n$ in the translation matrix will remain positive. By taking the value of the refractive index to be negative when a ray is travelling in the $-z$ direction, we can use the translation matrix throughout the analysis when reflecting surfaces are included in the optical system. Referring to Figure 2.17, the translation matrices between various planes are given as follows:

$$\mathscr{T}_{21} = \begin{pmatrix} 1 & \dfrac{d}{n} \\ 0 & 1 \end{pmatrix}$$

(2.4-31a)

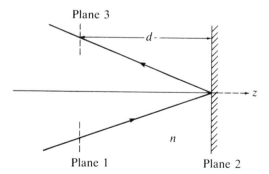

Figure 2.17 Rays reflected from a perfectly reflecting surface.

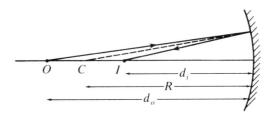

Figure 2.18 Reflection from a spherical mirror.

between planes 1 and 2;

$$\mathcal{T}_{32} = \begin{pmatrix} 1 & \dfrac{-d}{-n} \\ 0 & 1 \end{pmatrix} = \begin{pmatrix} 1 & \dfrac{d}{n} \\ 0 & 1 \end{pmatrix} \qquad (2.4\text{-}31\text{b})$$

between planes 2 and 3; and

$$\mathcal{T}_{31} = \begin{pmatrix} 1 & \dfrac{2d}{n} \\ 0 & 1 \end{pmatrix} \qquad (2.4\text{-}31\text{c})$$

between planes 1 and 3.

Similarly, we can modify the refraction matrix to describe the effects of a reflection from a surface (see Figure 2.18). All that is needed is to replace n_2 by $-n_1$ in the refraction power equation [Eq. (2.4-9b)],

$$p = \frac{n_2 - n_1}{R} = \frac{(-n_1) - n_1}{R} = \frac{-2n}{R}, \qquad (2.4\text{-}32)$$

where p is now the reflective power of the spherical mirror (reflector), R is the radius of curvature of the mirror, and $n = n_1$ is the refractive index for the medium in which the mirror is immersed. The *reflection matrix* $\tilde{\mathcal{R}}$ is given by

$$\tilde{\mathcal{R}} = \begin{pmatrix} 1 & 0 \\ -p & 1 \end{pmatrix} = \begin{pmatrix} 1 & 0 \\ \dfrac{2n}{R} & 1 \end{pmatrix}. \qquad (2.4\text{-}33)$$

The relationship between the image and object distances can be expressed once again by Eq. (2.4-19) with d_i replaced by $-d_i$ and $1/f$ replaced by $-2n/R$.

Example 2.9 Imaging by a Spherical Mirror

A point object is $d_o = 2$ m away from a concave mirror in air having a radius of curvature $R = -80$ cm. We want to find the position d_i of the image.

The matrix chain from the object to the image is

$$\mathscr{S} = \begin{pmatrix} 1 & d_i \\ 0 & 1 \end{pmatrix} \begin{pmatrix} 1 & 0 \\ \dfrac{2}{R} & 1 \end{pmatrix} \begin{pmatrix} 1 & d_o \\ 0 & 1 \end{pmatrix}$$

$$= \begin{pmatrix} 1 - 2.5d_i & 2 - 4d_i \\ -2.5 & -4 \end{pmatrix}, \tag{2.4-34}$$

and the image position is obtained by setting $\mathscr{B} = 2 - 4d_i = 0$. The image distance is positive, implying that the (real) image is formed on the left-hand side of the mirror.

Example 2.10 Optical Resonator

Consider an optical system consisting of two concave mirrors of radii of curvature R_1 and R_2, separated by a distance d, as shown in Figure 2.19. Such a system is called an optical *resonator* and forms an integral part of a laser system. For sustained oscillations, implying a constant laser output, the optical resonator must be stable.

To demonstrate the use of the matrix formalism for reflecting surfaces, we shall now obtain the condition for the resonator to be stable. In stable resonators, a ray must keep bounding back and forth and remain trapped inside in order that oscillations are sustained.

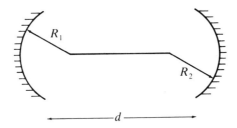

Figure 2.19 Schematic diagram of a resonator made up of two spherical mirrors.

Consider a ray starting at the left mirror. The system matrix chain describing the ray transformation by a round trip through the resonator is

$$\mathscr{S} = \begin{pmatrix} 1 & 0 \\ \dfrac{2}{R_1} & 0 \end{pmatrix} \begin{pmatrix} 1 & d \\ 0 & 1 \end{pmatrix} \begin{pmatrix} 1 & 0 \\ \dfrac{2}{R_2} & 1 \end{pmatrix} \begin{pmatrix} 1 & d \\ 0 & 1 \end{pmatrix}$$

$$= \begin{pmatrix} \mathscr{A} & \mathscr{B} \\ \mathscr{C} & \mathscr{D} \end{pmatrix}, \tag{2.4-35a}$$

where

$$\mathscr{A} = 1 + \frac{2d}{R_2},$$

$$\mathscr{B} = 2d\left(1 + \frac{d}{R_2}\right),$$

$$\mathscr{C} = 2\left\{\frac{1}{R_1} + \frac{1}{R_2}\left(1 + \frac{2d}{R_1}\right)\right\}, \tag{2.4-35b}$$

$$\mathscr{D} = \frac{2d}{R_1} + \left(1 + \frac{2d}{R_1}\right)\left(1 + \frac{2d}{R_2}\right).$$

The relation in Eq. (2.4-35) represents the matrix for one complete round trip of the ray. Thus, if $(x_1, \varkappa_1)$ represents the ray coordinates after one round trip, we can write

$$\begin{pmatrix} x_1 \\ \mathscr{v}_1 \end{pmatrix} = \begin{pmatrix} \mathscr{A} & \mathscr{B} \\ \mathscr{C} & \mathscr{D} \end{pmatrix} \begin{pmatrix} x_0 \\ \mathscr{v}_0 \end{pmatrix}, \qquad (2.4\text{-}36)$$

where $(x_0, \mathscr{v}_0)$ denotes the coordinates of a ray when it started from the left mirror. Hence, the coordinates of the ray $(x_m, \mathscr{v}_m)$ after m complete oscillations would be

$$\begin{pmatrix} x_m \\ \mathscr{v}_m \end{pmatrix} = \begin{pmatrix} \mathscr{A} & \mathscr{B} \\ \mathscr{C} & \mathscr{D} \end{pmatrix}^m \begin{pmatrix} x_0 \\ \mathscr{v}_0 \end{pmatrix}. \qquad (2.4\text{-}37)$$

For the resonator to be stable, it is required that the coordinates after m round trips through the resonator should not diverge as m increases. In order to obtain the stability criterion, we have to look at the mth power of the system matrix $\mathscr{S}$ given by Eq. (2.4-35). Noting that the system matrix has a determinant equal to 1, and defining an angle θ as

$$\cos\theta = \tfrac{1}{2}(\mathscr{A} + \mathscr{D}), \qquad (2.4\text{-}38)$$

we can show that

$$\begin{pmatrix} \mathscr{A} & \mathscr{B} \\ \mathscr{C} & \mathscr{D} \end{pmatrix}^m$$

$$= \frac{1}{\sin\theta} \begin{pmatrix} \mathscr{A}\sin m\theta - \sin(m-1)\theta & \mathscr{B}\sin m\theta \\ \mathscr{C}\sin m\theta & \mathscr{D}\sin m\theta - \sin(m-1)\theta \end{pmatrix}. \qquad (2.4\text{-}39)$$

Now, if θ is a complex number, the terms $\sin m\theta$ and $\sin(m-1)\theta$ in Eq. (2.4-39) diverge as $m \to \infty$. This happens if the magnitude of $\cos\theta$ is greater than 1. The preceding statements imply that for stability we must have

$$-1 \le \cos\theta \le 1 \qquad (2.4\text{-}40a)$$

or

$$-1 \le \tfrac{1}{2}(\mathscr{A}+ \mathscr{D}) \le 1. \qquad (2.4\text{-}40b)$$

Using Eq. (2.4-35b), Eqs. (2.4-40a) and (2.4-40b) become

$$0 \le \left(1 + \frac{d}{R_1}\right)\left(1 + \frac{d}{R_2}\right) \le 1, \qquad (2.4\text{-}41)$$

which is the stability criterion. This is alternatively written as

$$0 \le g_1 g_2 \le 1, \qquad (2.4\text{-}42)$$

where $g_1 = 1 + d/R_1$ and $g_2 = 1 + d/R_2$ are called the *g parameters* of the resonator. Figure 2.20 shows the stability diagram for optical resonators. Only those resonator configurations that lie in the shaded region correspond to a stable configuration. The point marked A corresponds to $R_1 = R_2 = \infty$, that is, $g_1 = g_2 = 1$, and the resonator is bounded by parallel plane mirrors. The point

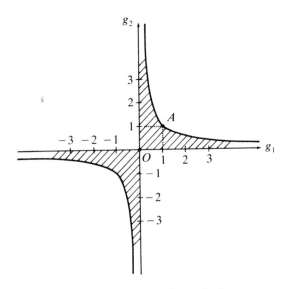

Figure 2.20 Stability diagram for optical resonators.

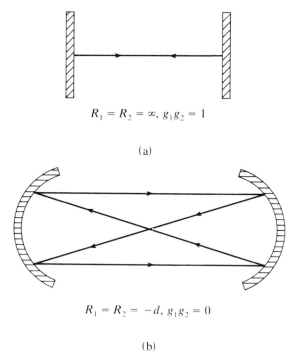

$R_1 = R_2 = \infty,\ g_1g_2 = 1$

(a)

$R_1 = R_2 = -d,\ g_1g_2 = 0$

(b)

Figure 2.21 Ray propagation inside resonator for (a) $R_1 = R_2 = \infty$ and (b) $R_1 = R_2 = d$.

marked O corresponds to a *confocal* configuration, where $R_1 = R_2 = -d$ or $g_1g_2 = 0$. Figures 2.21(a) and 2.21(b) show ray propagation inside these two resonators, respectively.

Problems

2.1 Derive the laws of reflection and refraction by considering the incident, reflected, and refracted light to comprise a stream of photons each characterized by a momentum $p = \hbar k$, where $\hbar = h/2\pi$ (h being Planck's constant), and k is the wave vector in the direction of ray propagation. Employ the law of conservation of momentum, assuming that the interface, say, y = constant, only affects the y component of the momentum. (This provides an alternative derivation of the laws of reflection and refraction.)

2.2 A commonly used construction for the refractive index variation in an optical fiber is given as

$$n(x) = n_o\left(1 - \Delta\left(\frac{x}{a}\right)^{\gamma}\right)^{1/2}.$$

Assume a ray as shown in Figure P2.2. Find the direction(s) of the ray at a distance $x = a$. Sketch the variation of the direction of the ray at $x = a$ as a function of Δ.

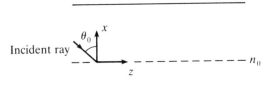

Figure P2.2

2.3 A thin glass beaker of 5-cm diameter is filled up with tap water of refraction index 1.4. Find the focal length of the cylindrical lens thus formed. Use the matrix formalism.

2.4 An object 3 cm tall is 15 cm in front of a converging lens of focal length 20 cm. Using the matrix formalism, find the location and magnification of the image. Draw a ray diagram from the object to the image.

2.5 Referring to Figure 2.15, verify Eqs. (2.4-23a) and (2.4-23b). Repeat the problem by interchanging L_1 and L_2.

2.6 Consider a system of two thin lenses, as shown in Figure 2.15. Let $f_1 = 4$ cm, $f_2 = -6$ cm, and $d = 3$ cm. An object 2 cm tall is placed 12 cm in front of the positive lens. Find the location and magnification of the image. Also, draw the complete ray diagram from the object to the final image.

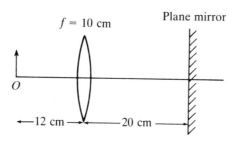

$f = 10$ cm Plane mirror

O

←—12 cm —→←——— 20 cm ———→

Figure P2.7

2.7 An object is placed 12 cm in front of a lens–mirror combination as shown in Figure P2.7. Using ray transfer matrix concepts, find the position and magnification of the image.

2.8 Prove Eq. (2.4-39) by mathematical induction.

2.9 Draw the ray propagation diagram similar to that shown in Figures 2.21(a) and 2.21(b) for the following parameters (where $d > 0$):
(a) $R_1 = \infty$ and $R_2 = -2d$ (hemispherical resonator);
(b) $R_1 = R_2 = -d/2$ (concentric resonator);
(c) $R_1 = d$ and $R_2 = \infty$.

References

2.1 Feynman, R., R. B. Leighton, and M. Sands (1963). *The Feynman Lectures on Physics*. Addison-Wesley, Reading, Massachusetts.

2.2 Gerard, A. and J. M. Burch (1975). *Introduction to Matrix Methods in Optics*. Wiley, New York.

2.3 Ghatak, A. K. (1980). *Optics*. Tata McGraw-Hill, New Delhi.

2.4 Goldstein, H. (1950). *Classical Mechanics*. Addison-Wesley, Reading, Massachusetts.

2.5 Hecht, E. and A. Zajac (1975). *Optics*. Addison-Wesley, Reading, Massachusetts.

2.6 Klein, M. V. (1970). *Optics*. Wiley, New York.

2.7 Nussbaum, A. and R. A. Phillips (1976). *Contemporary Optics for Scientists and Engineers*. Prentice-Hall, New York.

Chapter **3**   **Physical Optics**

In Chapter 2, we introduced some of the concepts of geometrical optics. However, as stated there, geometrical optics cannot account for wave effects such as diffraction. In this chapter, we introduce wave optics by starting from Maxwell's equations and deriving the wave equation. We thereafter discuss solutions of the wave equation and review power flow and polarization. We also study reflection and transmission across the interface between two semi-infinite media of different refractive indices and introduce the Fabry–Perot etalon. Having covered all the basics of electromagnetic theory, including dispersion, we discuss diffraction at length through use of the Fresnel diffraction formula, which is derived in a unique manner using Fourier transforms. In the process, we define the spatial transfer function and the impulse response of propagation. We also describe the distinguishing features of Fresnel and Fraunhofer diffraction and provide several illustrative examples. Specifically, we analyze diffraction from rectangular and circular apertures, introduce the Fresnel zone plate, and analyze the diffraction of a Gaussian beam. In all cases, we restrict ourselves to propagation in a medium with a constant refractive index. For completeness, we also discuss wave propagation through a medium having a refractive index profile, as in an optical fiber.

3.1 Maxwell's Equations: A Review

In the study of electromagnetics, we are concerned with four vector quantities called electromagnetic fields: the electric field strength $\mathbf{E}$ (V/m); the electric flux density $\mathbf{D}$ (C/m^2); the magnetic field strength $\mathbf{H}$ (A/m); and the magnetic flux density $\mathbf{B}$ (Wb/m^2). The fundamental theory of electromagnetic fields is based on *Maxwell's equations*. In differential form, these are expressed as

$$\nabla \cdot \mathbf{D} = \rho, \tag{3.1-1}$$

$$\nabla \cdot \mathbf{B} = 0, \tag{3.1-2}$$

$$\nabla \times \mathbf{E} = -\frac{\partial \mathbf{B}}{\partial t}, \tag{3.1-3}$$

$$\nabla \times \mathbf{H} = \mathbf{J} = \mathbf{J}_c + \frac{\partial \mathbf{D}}{\partial t}, \tag{3.1-4}$$

where $\mathbf{J}$ is the current density (A/m^2) and ρ denotes the electric charge density (C/m^3). $\mathbf{J}_c$ and ρ are the sources generating the electromagnetic fields.

We can summarize the physical interpretation of Maxwell's equations as follows:

Equation (3.1-1) is the differential representation of *Gauss's law for electric fields*. To convert this to an integral form, which is more physically transparent, we integrate Eq. (3.1-1) over a volume V bounded by a surface S and use the *divergence theorem* (or *Gauss's theorem*),

$$\int_V \nabla \cdot \mathbf{D} \, dV = \oint_S \mathbf{D} \cdot d\mathbf{S}, \tag{3.1-5}$$

to get

$$\oint_S \mathbf{D} \cdot d\mathbf{S} = \int_V \rho \, dV. \tag{3.1-6}$$

This states that the *electric flux $\oint_S \mathbf{D} \cdot d\mathbf{S}$ flowing out of a surface S enclosing a volume V equals the total charge enclosed in the volume.*

Equation (3.1-2) is the *magnetic analog* of Eq. (3.1-1) and can be converted to an integral form similar to Eq. (3.1-6) by using the divergence theorem once again,

$$\oint_S \mathbf{B} \cdot dS = 0. \tag{3.1-7}$$

The right-hand sides (RHSs) of Eqs. (3.1-2) and (3.1-7) are zero because, in the classical sense, magnetic monopoles do not exist. Thus, the *magnetic flux is always conserved*.

Equation (3.1-3) enunciates *Faraday's law of induction*. To convert this to an integral form, we integrate over an open surface S bounded by a line C and use *Stokes's theorem*,

$$\int_S (\nabla \times \mathbf{E}) \cdot dS = \oint_C \mathbf{E} \cdot dl, \tag{3.1-8}$$

to get

$$\oint_C \mathbf{E} \cdot dl = -\int_S \frac{\partial \mathbf{B}}{\partial t} \cdot dS. \tag{3.1-9}$$

This states that the *electromotive force* (emf) $\oint_C \mathbf{E} \cdot dl$ *induced in a loop is equal to the time rate of change of the magnetic flux passing through the area of the loop*. The emf is induced in a sense such that it opposes the variation of the magnetic field, as indicated by the minus sign in Eq. (3.1-9); this is known as *Lenz's law*.

Analogously, the integral form of Eq. (3.1-4) reads

$$\oint_C \mathbf{H} \cdot dl = \int_S \frac{\partial \mathbf{D}}{\partial t} \cdot dS + \int_S \mathbf{J}_c \cdot dS, \tag{3.1-10}$$

which states that the *line integral* of $\mathbf{H}$ *around a closed loop C equals the total current (conduction and displacement) passing through the surface of the loop*. When first formulated by Ampere, Eqs. (3.1-4) and (3.1-10) only had the conduction current term $\mathbf{J}_c$ on the RHS. Maxwell proposed the addition of the displacement current term $\partial \mathbf{D}/\partial t$ to include the effect of currents flowing through, for instance, a capacitor.

For a given current and charge density distribution, note that there are four equations [Eqs. (3.1-1)–(3.1-4)] and, at first sight, four unknowns that need to be determined to solve a given electromagnetic problem. As such, the problem appears well-posed. However, a closer examination reveals that Eqs. (3.1-3) and (3.1-4), which are vector equations, are really equivalent to six scalar equations. Also, by virtue of the *continuity equation*,

$$\nabla \cdot \mathbf{J}_c + \frac{\partial \rho}{\partial t} = 0, \tag{3.1-11}$$

Eq. (3.1-1) is not independent of Eq. (3.1-4) and, similarly, Eq. (3.1-2) is a consequence of Eq. (3.1-3). We can verify this by taking the divergence on both sides of Eqs. (3.1-3) and (3.1-4) and by using the continuity equation [Eq. (3.1-11)] and a vector relation,

$$\nabla \cdot (\nabla \times \mathbf{A}) = 0, \tag{3.1-12}$$

to simplify. The upshot of this discussion is that, strictly speaking, there are six independent scalar equations and twelve unknowns (viz., the x, y, and z components of $\mathbf{E}$, $\mathbf{D}$, $\mathbf{H}$, and $\mathbf{B}$) to solve for. The six more scalar equations required are provided by the *constitutive relations*,

$$\mathbf{D} = \epsilon \mathbf{E}, \tag{3.1-13a}$$

$$\mathbf{B} = \mu \mathbf{H}, \tag{3.1-13b}$$

where ϵ denotes the permittivity (F/m) and μ the permeability (H/m) of the medium. Note that we have written ϵ and μ as scalar constants. This is true for a *linear, homogeneous, isotropic* medium. A medium is *linear* if its properties do not depend on the amplitude of the fields in the medium. It is *homogeneous* if its properties are not functions of space. The medium is, furthermore, *isotropic* if its properties are the same in all direction from any given point. For most of this book, we will assume the medium to be linear, homogeneous, and isotropic. However, anisotropic materials will be studied when we examine electrooptic effects in Chapter 6, and the propagation of waves in nonlinear media will be examined in Chapter 8. The coverage of inhomogeneous materials like optical fibers will appear in Section 3.6.

Returning our focus to linear, homogeneous, isotropic media, constants worth remembering are the values of ϵ and μ for free space or vacuum: $\epsilon_0 = (1/36\pi) \times 10^{-9}$ F/m and $\mu_0 = 4\pi \times 10^{-/}$ H/m. For *dielectrics*, the value of ϵ is greater than ϵ_0, because the **D** field is composed of a free-space part $\epsilon_0 \mathbf{E}$ and a material part characterized by a *dipole moment density* **P** (C/m^2). **P** is related to the electric field **E** as

$$\mathbf{P} = \chi \epsilon_0 \mathbf{E}, \qquad (3.1\text{-}14)$$

where χ is the *electric susceptibility* and indicates the ability of the electric dipoles in the dielectric to align themselves with the electric field. The **D** field is the sum of $\epsilon_0 \mathbf{E}$ and **P**,

$$\mathbf{D} = \epsilon_0 \mathbf{E} + \mathbf{P} = \epsilon_0(1 + \chi)\mathbf{E} \triangleq \epsilon_0 \epsilon_r \mathbf{E}, \qquad (3.1\text{-}15)$$

where ϵ_r is the relative permittivity, so that

$$\epsilon = (1 + \chi)\epsilon_0. \qquad (3.1\text{-}16)$$

Similarly, for magnetic materials μ is greater than μ_0.

3.2 Linear Wave Propagation

In this section, we first derive the wave equation and review some of the travelling-wave type solutions of the equation in different coordinate systems. We define the concept of intrinsic impedance, Poynting vector and intensity, and introduce the subject of polarization. Thereafter, we analyze plane-wave propagation across the interface between two semiinfinite media of different refractive indices and state the ramifications of the results in connection with the reflection of a beam under certain conditions. We also introduce the Fabry–Perot etalon and discuss the rudiments of dispersion and its effects on wave propagation. Finally, we illustrate another method to analyze the solutions of the homogeneous wave equation by deriving the eikonal equations, which are common in optics and which provide the link between ray and wave optics. The latter discussion also establishes the link with the following section, where we will study the effects of diffraction at length.

3.2.1 Travelling-Wave Solutions

In Section 3.1 we enunciated Maxwell's equations and the constitutive relations. For a given $\mathbf{J}_c$ and ρ, we remarked that we could, in fact, solve for the components of the electric field $\mathbf{E}$. In this subsection, we see how this can be done. We derive the wave equation describing the propagation of the electric and magnetic fields and find its general solutions in different coordinate systems. By taking the curl of both sides of Eq. (3.1-3) we have

$$\boldsymbol{\nabla} \times \boldsymbol{\nabla} \times \mathbf{E} = -\boldsymbol{\nabla} \times \frac{\partial \mathbf{B}}{\partial t} = -\frac{\partial}{\partial t}(\boldsymbol{\nabla} \times \mathbf{B}) = -\mu \frac{\partial}{\partial t}(\boldsymbol{\nabla} \times \mathbf{H}),$$

(3.2-1)

where we have used the second of the constitutive relations [Eq. (3.1-13b)] and assumed μ to be space- and time-independent. Now, employing Eq. (3.1-4), Eq. (3.2-1) becomes

$$\boldsymbol{\nabla} \times \boldsymbol{\nabla} \times \mathbf{E} = -\mu\epsilon \frac{\partial^2 \mathbf{E}}{\partial t^2} - \mu \frac{\partial \mathbf{J}_c}{\partial t},$$

(3.2-2)

where we have used the first of the constitutive relations [Eq. (3.1-13a)] and assumed ϵ to be time-independent. Then, by using the vector relationship

$$\boldsymbol{\nabla} \times \boldsymbol{\nabla} \times \mathbf{A} = \boldsymbol{\nabla}(\boldsymbol{\nabla} \cdot \mathbf{A}) - \nabla^2 \mathbf{A}, \qquad \nabla^2 = \boldsymbol{\nabla} \cdot \boldsymbol{\nabla}, \quad (3.2\text{-}3)$$

in Eq. (3.2-2), we get

$$\nabla^2 \mathbf{E} - \mu\epsilon \frac{\partial^2 \mathbf{E}}{\partial t^2} = \mu \frac{\partial \mathbf{J}_c}{\partial t} + \boldsymbol{\nabla}(\boldsymbol{\nabla} \cdot \mathbf{E}).$$

(3.2-4)

If we now assume the permittivity ϵ to be space-independent as well, then we can recast the first of Maxwell's equations [Eq. (3.1-1)] in the form

$$\boldsymbol{\nabla} \cdot \mathbf{E} = \frac{\rho}{\epsilon},$$

(3.2-5)

using the first of the constitutive relations [Eq. (3.1-13a)]. Incorporating Eq. (3.2-5) into Eq. (3.2-4), we finally obtain

$$\nabla^2 \mathbf{E} - \mu\epsilon \frac{\partial^2 \mathbf{E}}{\partial t^2} = \mu \frac{\partial \mathbf{J}_c}{\partial t} + \frac{1}{\epsilon}\boldsymbol{\nabla}\rho,$$

(3.2-6)

which is a wave equation having source terms on the RHS. In fact, Eq. (3.2-6), being a vector equation, is really equivalent to three scalar equations, one for every component of **E**. Expressions for the Laplacian (∇^2) operator in Cartesian (x, y, z), cylindrical (r, ϕ, z), and spherical (R, θ, ϕ) coordinates are given as follows:

$$\nabla^2_{\text{rect}} = \frac{\partial^2}{\partial x^2} + \frac{\partial^2}{\partial y^2} + \frac{\partial^2}{\partial z^2}; \tag{3.2-7}$$

$$\nabla^2_{\text{cyl}} = \frac{\partial^2}{\partial r^2} + \frac{1}{r}\frac{\partial}{\partial r} + \frac{1}{r^2}\frac{\partial^2}{\partial \phi^2} + \frac{\partial^2}{\partial z^2}; \tag{3.2-8}$$

$$\nabla^2_{\text{sph}} = \frac{\partial^2}{\partial R^2} + \frac{2}{R}\frac{\partial}{\partial R} + \frac{1}{R^2\sin^2\theta}\frac{\partial^2}{\partial \phi^2} + \frac{1}{R^2}\frac{\partial^2}{\partial \theta^2} + \frac{\cot\theta}{R^2}\frac{\partial}{\partial \theta}. \tag{3.2-9}$$

In space free of all sources ($J_c = 0$, $\rho = 0$), Eq. (3.2-6) reduces to the *homogeneous wave equation*

$$\nabla^2 \mathbf{E} = \mu\epsilon \frac{\partial^2 \mathbf{E}}{\partial t^2}. \tag{3.2-10}$$

A similar equation may be derived for the magnetic field **H**,

$$\nabla^2 \mathbf{H} = \mu\epsilon \frac{\partial^2 \mathbf{H}}{\partial t^2}. \tag{3.2-11}$$

We caution readers that the ∇^2 operator, as written in Eqs. (3.2-7)–(3.2-9), must be applied only after decomposing Eqs. (3.2-10) and (3.2-11) into scalar equations for three orthogonal components in **a**x, **a**y and **a**z. However, for the rectangular coordinate case only, these scalar equations may be recombined and interpreted as the Laplacian ∇^2_{rect} acting on the total vector.

Note that the quantity $\mu\epsilon$ has the units of $(1/\text{velocity})^2$. We call this velocity v and define it as

$$v^2 = \frac{1}{\mu\epsilon}. \tag{3.2-12}$$

For free space, $\mu = \mu_0$, $\epsilon = \epsilon_0$, and $v = c$. We can calculate the value of c from the values of ϵ_0 and μ_0 mentioned in Section 3.1. This works out to 3×10^8 m/s. This theoretical value, first calculated by Maxwell, was in remarkable agreement with Fizeau's previously measured speed of light (315,300 km/s). This led Maxwell to conclude that light is *an electromagnetic disturbance in the form of waves propagated through the electromagnetic field according to electromagnetic laws.*

Let us now examine the solutions of equations of the type of Eqs. (3.2-10) or (3.2-11) in different coordinate systems. For simplicity, we will analyze the homogeneous wave equation

$$\frac{\partial^2 \psi}{\partial t^2} - v^2 \nabla^2 \psi = 0, \qquad (3.2\text{-}13)$$

where ψ may represent a component of the electric field $\mathbf{E}$ or of the magnetic field $\mathbf{H}$ and where v is the velocity of the wave.

In Cartesian coordinates, the general solution is

$$\psi(x, y, z, t) = c_1 f(\omega_0 t - k_{0x} x - k_{0y} y - k_{0z} z)$$

$$+ c_2 g(\omega_0 t + k_{0x} x + k_{0y} y + k_{0z} z),$$

$$c_1, c_2 \text{ constants}, \quad (3.2\text{-}14)$$

with the condition

$$\frac{\omega_0^2}{k_{0x}^2 + k_{0y}^2 + k_{0z}^2} \triangleq \frac{\omega_0^2}{k_0^2} = v^2. \qquad (3.2\text{-}15)$$

In Eq. (3.2-15), ω_0 is the (*angular*) *frequency* (rad/s) of the wave and k_0 is the *propagation constant* (rad/m) in the medium. Since the ratio ω_0/k_0 is a constant, the medium of propagation is said to be *nondispersive*. [We will discuss dispersion formally in Section 3.2.4.] We can then reexpress Eq. (3.2-14) as

$$\psi(x, y, z, t) = c_1 f(\omega_0 t - \mathbf{k}_0 \cdot \mathbf{R}) + c_2 g(\omega_0 t + \mathbf{k}_0 \cdot \mathbf{R}), \quad (3.2\text{-}16)$$

where

$$\mathbf{R} = x\mathbf{a}_x + y\mathbf{a}_y + z\mathbf{a}_z, \qquad (3.2\text{-}17a)$$

$$\mathbf{k}_0 \triangleq k_{0x}\mathbf{a}_x + k_{0y}\mathbf{a}_y + k_{0z}\mathbf{a}_z. \qquad (3.2\text{-}17b)$$

$\mathbf{k}_0$ is called the *propagation vector* and $|\mathbf{k}_0| = k_0$; $\mathbf{a}_x$, $\mathbf{a}_y$, and $\mathbf{a}_z$ denote the unit vectors in the x, y, and z directions, respectively.

In one spatial dimension (viz., z), the wave equation [Eq. (3.2-13)] reads

$$\frac{\partial^2 \psi}{\partial t^2} - v^2 \frac{\partial^2 \psi}{\partial z^2} = 0, \qquad (3.2\text{-}18)$$

and its general solution is

$$\psi(z,t) = c_1 f(\omega_0 t - k_o z) + c_2 g(\omega_0 t + k_o z), \qquad v = \frac{\omega_0}{k_0}. \tag{3.2-19}$$

Note that Eq. (3.2-14) or (3.2-16) comprises the superposition of two waves, travelling in opposite directions. We can define a *wave* as a disturbance of some form characterized by a recognizable amplitude and a recognizable velocity of propagation. Although this definition sounds rather loose at the moment, we will see that it perfectly describes all the different types of waves that we will encounter in this book. Let us now consider a special case: $c_1 \neq 0$, $c_2 = 0$. Observe that if ψ is a constant, so is $\omega_0 t - \mathbf{k}_0 \cdot \mathbf{R}$. Hence,

$$\mathbf{k}_0 \cdot \mathbf{R} = \omega_0 t + \text{constant}. \qquad (3.2\text{-}20)$$

But this is the equation of a plane perpendicular to $\mathbf{k}_0$ with t as a parameter; hence the wave is called a *plane wave*. With increasing t, $\mathbf{k}_0 \cdot \mathbf{R}$ must increase so that Eq. (3.2-20) always holds. For instance, if $\mathbf{k}_0 = k_0 \mathbf{a}_z$ ($k_0 > 0$) and $\mathbf{R} = z\mathbf{a}_z$, z must increase as t increases. This means that the wave propagates in the $+z$ direction. For $c_1 = 0$, $c_2 \neq 0$, we have a plane wave travelling in the opposite direction. The *wavefronts*, defined as the surfaces joining all points of equal phase $\omega_0 t \pm \mathbf{k}_0 \cdot \mathbf{R}$, are planar.

Consider now the cylindrical coordinate system. The simplest case is that of cylindrical symmetry, which requires that $\psi(r, \phi, z, t) = \psi(r, z, t)$. The ϕ-independence means that a plane perpendicular to the z axis will intersect the wavefront in a circle. Even in this very simple case, no solutions in terms of arbitrary functions can be found as was done previously for plane waves. However, we can show that harmonic, z- and ϕ-independent solutions of the form

$$\psi(r, t) \sim \frac{C}{r^{1/2}} \exp[j(\omega_0 t \pm k_0 r)], \qquad r \gg 0 \text{ and } C \text{ constant},$$

$$(3.2\text{-}21)$$

approximately satisfy the wave equation [Eq. (3.2-13)]. We remark that the exact solution has a Bessel-function type dependence on r if we assume ψ to be *time-harmonic*, that is, of the form $\psi = \text{Re}[\psi_p(r)\exp j\omega_0 t]$, where $\text{Re}[\cdot]$ means "the real part of."

Finally, we present solutions of the wave equation in a spherical coordinate system. For spherical symmetry ($\partial/\partial\phi = 0 = \partial/\partial\theta$), the wave equation, Eq. (3.2-13), with Eq. (3.2-9) assumes the form

$$R\left(\frac{\partial^2 \psi}{\partial R^2} + \frac{2}{R}\frac{\partial \psi}{\partial R}\right) = \frac{\partial^2(R\psi)}{\partial R^2} = \frac{1}{v^2}\frac{\partial^2(R\psi)}{\partial t^2}. \qquad (3.2\text{-}22)$$

Now, Eq. (3.2-22) is of the same form as Eq. (3.2-18). Hence, using Eq. (3.2-19), we can write down the solution of Eq. (3.2-22) as

$$\psi = \frac{c_1}{R}f(\omega_0 t - k_0 R) + \frac{c_2}{R}g(\omega_0 t + k_0 R), \qquad c_1, c_2 \text{ constants},$$

$$(3.2\text{-}23)$$

with $\omega_0/k_0 = v$. The wavefronts are spherical, defined by

$$k_0 R = k_0[x^2 + y^2 + z^2]^{1/2} = \omega_0 t + \text{constant}. \qquad (3.2\text{-}24)$$

3.2.2 Intrinsic Impedance, the Poynting Vector, and Polarization

So far in our discussion of Maxwell's equations and the wave equation and its solutions, we made no comments on the components of **E** and **H**. The solutions of the wave equation [Eq. (3.2-13)] in different coordinate systems are valid for every component of **E** and **H**. We point out here that the solutions of the wave equation discussed previously hold, in general, only in an unbounded medium. We will discuss the nature of the solutions in the presence of specific boundary conditions in Section 3.2.3.

In this subsection, we first show that electromagnetic wave propagation is transverse in nature in an unbounded medium and derive the relationships between the existing electric and magnetic fields. In this connection, we define the intrinsic or characteristic impedance of a medium, which is similar in concept to the characteristic impedance of a transmission line. We also introduce the concept of power flow during electromagnetic propagation and define the Poynting vector and the irradiance. Also, we expose readers to the different types of polarization that the electric field might possess during propagation.

In an unbounded isotropic, linear, homogeneous medium free of sources, electromagnetic wave propagation is *transverse* in nature. This means that the only components of **E** and **H** are those that are transverse to the direction of propagation. To check this, we consider propagating electric and magnetic fields of the forms

$$\mathbf{E} = \mathbf{E}_x + \mathbf{E}_y + \mathbf{E}_z$$

$$\triangleq \text{Re}\{E_{0x} \exp[j(\omega_0 t - k_0 z)]\mathbf{a}_x$$

$$+ E_{0y} \exp[j(\omega_0 t - k_0 z)]\mathbf{a}_y + E_{0z} \exp[j(\omega_0 t - k_0 z)]\mathbf{a}_z\},$$

$$(3.2\text{-}25)$$

$$\mathbf{H} = \mathbf{H}_x + \mathbf{H}_y + \mathbf{H}_z$$

$$\triangleq \text{Re}\{H_{0x} \exp[j(\omega_0 t - k_0 z)]\mathbf{a}_x$$

$$+ H_{0y} \exp[j(\omega_0 t - k_0 z)]\mathbf{a}_y + H_{0z} \exp[j(\omega_0 t - k_0 z)]\mathbf{a}_z\},$$

$$(3.2\text{-}26)$$

where E_{0x}, E_{0y}, E_{0z} and H_{0x}, H_{0y}, H_{0z} are (complex) constants. We put Eq. (3.2-25) in the first of Maxwell's equations [i.e., Eq. (3.1-1)], with $\rho = 0$, and invoke the constitutive relation, Eq. (3.1-13a), to derive

$$\frac{\partial}{\partial z}\{E_{0z}\exp[j(\omega_0 t - k_0 z)]\} = 0,$$

implying

$$E_{0z} = 0. \tag{3.2-27a}$$

This means that *there is **no** component of the electric field in the direction of propagation.* The only possible components of **E** then must be in a plane transverse to the direction of propagation. Similarly, using Eqs. (3.1-2) and (3.1-13b) we can show that

$$H_{0z} = 0. \tag{3.2-27b}$$

Furthermore, substitution of Eqs. (3.2-25) and (3.2-26) with $E_{0z} = 0 = H_{0z}$ into the third of Maxwell's equations, Eq. (3.1-3), yields

$$k_0 E_{0y}\mathbf{a}_x - k_0 E_{0x}\mathbf{a}_y = -\mu\omega_0(H_{0x}\mathbf{a}_x + H_{0y}\mathbf{a}_y).$$

We can then write [using Eqs. (3.2-12) and (3.2-15)]

$$H_{0x} = -\frac{1}{\eta}E_{0y}, \qquad H_{0y} = \frac{1}{\eta}E_{0x}, \tag{3.2-28}$$

where

$$\eta \triangleq \frac{\omega_0}{k_0}\mu = \upsilon\mu = \left(\frac{\mu}{\epsilon}\right)^{1/2} \tag{3.2-29}$$

is called the *intrinsic* or *characteristic impedance* of the medium. The characteristic impedance has the units of V/A, or Ω. Its value for free space is $\eta_0 = 377\ \Omega$. Now, using Eqs. (3.2-25)–(3.2-29), we see that

$$\mathbf{E} \cdot \mathbf{H} = 0, \tag{3.2-30}$$

meaning that the *electric and magnetic fields are orthogonal to each other*, and that $\mathbf{E} \times \mathbf{H}$ is along the direction of propagation (z) of the electromagnetic field. Similar relationships can be established in other coordinate systems.

Note that $\mathbf{E} \times \mathbf{H}$ has the units of W/m^2, reminiscent of power per unit area. All electromagnetic waves carry energy, and for isotropic media the energy flow occurs in the direction of propagation of the wave. As we shall see in Chapter 6, this is not true for anisotropic media. The *Poynting vector* $\mathbf{S}$, defined as

$$\mathbf{S} = \mathbf{E} \times \mathbf{H}, \qquad (3.2\text{-}31)$$

is a power density vector associated with an electromagnetic field.

In a linear, homogeneous, isotropic unbounded medium, we can choose the electric and magnetic fields to be of the form

$$\mathbf{E}(z, t) = \text{Re}\left[E_0 \exp[j(\omega_0 t - k_0 z)]\right]\mathbf{a}_x, \qquad (3.2\text{-}32a)$$

$$\mathbf{H}(z, t) = \text{Re}\left[\frac{E_0}{\eta} \exp[j(\omega_0 t - k_0 z)]\right]\mathbf{a}_y, \qquad (3.2\text{-}32b)$$

where E_0 is, in general, a complex quantity. This choice is consistent with Eqs. (3.2-28)–(3.2-30). If we find $\mathbf{S}$, note that it is a function of time. It is more convenient, therefore, to define the time-averaged power, or *irradiance I*,

$$I = \langle |\mathbf{S}| \rangle = \frac{\omega_0}{2\pi} \int_0^{2\pi/\omega_0} |\mathbf{S}| \, dt = \frac{|E_0|^2}{2\eta} = \epsilon v \frac{|E_0|^2}{2}. \quad (3.2\text{-}33)$$

In the chapters on optical information processing (Chapters 4 and 5), the irradiance will be referred to as the *intensity*. Unless otherwise stated, it will always be taken to be *proportional to* the magnitude squared of the complex field.

In the remainder of this subsection, we introduce readers to the concept of *polarization* of the electric field. The polarization describes the time-varying behavior of the electric field vector at a given point in space. A separate description of the magnetic field is not necessary, because the direction of $\mathbf{H}$ is definitely related to that of $\mathbf{E}$.

Assume, for instance, that in Eq. (3.2-25), $E_{0z} = 0$ and

$$E_{0x} = |E_{0x}|, \qquad E_{0y} \stackrel{*}{=} |E_{0y}|e^{-j\phi_0}, \qquad (3.2\text{-}34)$$

where ϕ_0 is a constant.

First, consider the case where $\phi_0 = 0$ or $\pm\pi$. Then, the two components of **E** are in phase, and

$$\mathbf{E} = \left(|E_{0x}|\mathbf{a}_x \pm |E_{0y}|\mathbf{a}_y\right)\cos(\omega_0 t - k_0 z). \qquad (3.2\text{-}35)$$

The direction of **E** is fixed on a plane perpendicular to the direction of propagation (this plane is referred to as the *plane of polarization*) and does not vary with time, and the electric field is said to be *linearly polarized*.

As a second case, assume $\phi_0 = \pm\pi/2$ and $|E_{0x}| = |E_{0y}| = E_0$. In this case, from Eq. (3.2-25),

$$\mathbf{E} = E_0 \cos(\omega_0 t - k_0 z)\mathbf{a}_x \pm E_0 \sin(\omega_0 t - k_0 z)\mathbf{a}_y. \quad (3.2\text{-}36)$$

When monitored at a certain point $z = z_0$ during propagation, the direction of **E** is no longer fixed along a line, but varies with time according to $\theta = \omega_0 t - k_0 z_0$, where θ represents the angle between **E** and the (transverse) x axis. The amplitude of **E** (which is equal to E_0) is, however, still a constant. This is an example of *circular polarization* of the electric field. When $\phi_0 = -\pi/2$, $\mathbf{E}_y$ leads $\mathbf{E}_x$ by $\pi/2$ [see Eq. (3.2-36)]. Hence, as a function of time, **E** describes a clockwise circle in the x–y plane as seen head-on at $z = z_0$. Similarly, for $\phi_0 = +\pi/2$, **E** describes a counterclockwise circle.

In the general case,

$$\mathbf{E} = \mathbf{E}_x + \mathbf{E}_y \triangleq E_x\mathbf{a}_x + E_y\mathbf{a}_y$$

$$= |E_{0x}|\cos(\omega_0 t - k_0 z)\mathbf{a}_x + |E_{0y}|\cos(\omega_0 t - k_0 z - \phi_0)\mathbf{a}_y. \quad (3.2\text{-}37)$$

As in the case of circularly polarized waves (where $|E_{0x}|^2 + |E_{0y}|^2 = E_0^2 = $ constant), the direction of **E** is no longer fixed because E_x and E_y vary with time. We can trace this variation on the E_x–E_y plane by eliminating the harmonic variations of $(\omega_0 t - k_0 z)$ in

Eq. (3.2-37). To do this, we note that

$$\frac{E_y}{|E_{0y}|} = \cos(\omega_0 t - k_0 z)\cos \phi_0 + \sin(\omega_0 t - k_0 z)\sin \phi_0$$

$$= \frac{E_x}{|E_{0x}|} \cos \phi_0 + \left\{ 1 - \left(\frac{E_x}{|E_{0x}|} \right)^2 \right\}^{1/2} \sin \phi_0. \quad (3.2\text{-}38)$$

After a little algebra we can reexpress Eq. (3.2-38) as

$$\left(\frac{E_x}{|E_{0x}|} \right)^2 - 2\left(\frac{E_x}{|E_{0x}|} \right)\left(\frac{E_y}{|E_{0y}|} \right)\cos \phi_0 + \left(\frac{E_y}{|E_{0y}|} \right)^2 = \sin^2 \phi_0, \quad (3.2\text{-}39)$$

which is the equation of an ellipse; hence, the wave is said to be *elliptically polarized*. Note that for values of ϕ_0 equal to 0 or $\pm\pi$ and $\pm\pi/2$ (with $|E_{0x}| = |E_{0y}| = E_0$), the polarization configurations reduce to the linearly and circularly polarized cases, respectively.

Figure 3.1 illustrates various polarization configurations corresponding to different values of ϕ_0 to demonstrate clearly linear, circular, and elliptical polarizations. In this figure, we show the direction of rotation of the **E**-field vector with time, and its magnitude for various ϕ_0. When $\phi = 0$ or $\pm\pi$, the E field is linearly polarized and the **E** vector does not rotate. Unless otherwise stated, we will, throughout the book, assume all electric fields to be linearly

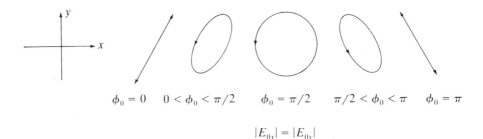

$$\phi_0 = 0 \quad 0 < \phi_0 < \pi/2 \quad \phi_0 = \pi/2 \quad \pi/2 < \phi_0 < \pi \quad \phi_0 = \pi$$

$$|E_{0x}| = |E_{0y}|$$

Figure 3.1 Various polarization configurations corresponding to different values of ϕ_0. ($|E_{0x}| \neq |E_{0y}|$, unless otherwise stated.)

polarized and choose the transverse axes such that one of them coincides with the direction of the electric field. The magnetic field will, therefore, be along the other transverse axis and will be related to the electric field via the characteristic impedance η.

The state of polarization of light travelling in the z direction can alternately be represented in terms of a complex vector $(E_{0x}, E_{0y})^T$. For instance, $(1\ 0)^T$, $(0\ 1)^T$, and $(\frac{1}{2})^{1/2}\ (1\ 1)^T$ are examples of linearly polarized light. Such vectors are called *Jones vectors*. Furthermore, the transformation from one polarization state to another may be represented by a matrix called the *Jones matrix*. For instance, transformation from $(1\ 0)^T$ to $(0\ 1)^T$ may be realized by applying the matrix $\begin{pmatrix} 0 & 1 \\ 1 & 0 \end{pmatrix}$. An optical element that can perform this transformation is the *half-wave plate*. This is discussed in Chapter 6.

3.2.3 Electromagnetic Boundary Conditions and Fresnel Equations

So far we have studied wave propagation through an unbounded medium. In this subsection, we discuss wave propagation through two semi-infinite media sharing a common interface. To this end, we first state the electromagnetic boundary conditions. We then investigate the effects of polarization on the reflection and transmission of electromagnetic waves at the interface between two linear isotropic dielectrics and derive the Fresnel equations. We also include a discussion on total internal reflection and establish the properties of evanescent waves.

We state, in words, the *electromagnetic boundary conditions* at the interface between two linear, homogeneous, isotropic media with no free charges or currents at the interface. *The tangential components of E and H are continuous across an interface. Also, the normal components of D and B are continuous.*

Now consider a plane polarized wave incident on the interface of two dielectrics characterized by refractive indices n_1 and n_2 defined as

$$n_1 = \frac{c}{v_1} = \left[\frac{\epsilon_1}{\epsilon_0}\right]^{1/2}, \qquad n_2 = \frac{c}{v_2} = \left[\frac{\epsilon_2}{\epsilon_0}\right]^{1/2}. \qquad (3.2\text{-}40)$$

Let

$$\mathbf{E}_i = \mathbf{E}_{i0} \cos(\omega_i t - \mathbf{k}_i \cdot \mathbf{R}), \tag{3.2-41a}$$

$$\mathbf{E}_r = \mathbf{E}_{r0} \cos(\omega_r t - \mathbf{k}_r \cdot \mathbf{R} - \phi_r), \tag{3.2-41b}$$

and

$$\mathbf{E}_t = \mathbf{E}_{t0} \cos(\omega_t t - \mathbf{k}_t \cdot \mathbf{R} - \phi_t) \tag{3.2-41c}$$

denote the incident, reflected, and transmitted fields, respectively. By employing the electromagnetic boundary conditions, we can show that the temporal frequencies of the three fields are equal. This is also intuitively transparent because the response of the electrons executing forced vibrations in both media must have the same frequency as that of the excitation. The laws of reflection and refraction as derived from ray optics in Chapter 2 [see Eqs. (2.2-3b) and (2.2-6b)] follows directly as a consequence of matching the boundary conditions at the interface. We can also show that all the **k** vectors are coplanar. This plane is called the *plane of incidence*.

We now investigate the effects of polarization on the reflection and transmission properties of an interface between two linear isotropic nonmagnetic dielectric media. To this end, note that we can decompose the incident, reflected, and transmitted electric fields at the interface into two orthogonal components, one parallel and the other perpendicular to the plane of incidence, and treat each of these constituents separately.

Parallel Polarization

In this case, the **E** vectors of all three waves are in the plane of incidence, as shown in Figure 3.2. We assume the forms of the three propagating electric and magnetic fields as

$$\mathbf{E}_i = \mathbf{E}_{i0} \cos(\omega_0 t - \mathbf{k}_i \cdot \mathbf{R}), \quad \mathbf{H}_i = \mathbf{H}_{i0} \cos(\omega_0 t - \mathbf{k}_i \cdot \mathbf{R}),$$

$$\mathbf{E}_r = \mathbf{E}_{r0} \cos(\omega_0 t - \mathbf{k}_r \cdot \mathbf{R}), \quad \mathbf{H}_r = \mathbf{H}_{r0} \cos(\omega_0 t - \mathbf{k}_r \cdot \mathbf{R}),$$
$$\tag{3.2-42}$$

$$\mathbf{E}_t = \mathbf{E}_{t0} \cos(\omega_0 t - \mathbf{k}_t \cdot \mathbf{R}), \quad \mathbf{H}_t = \mathbf{H}_{t0} \cos(\omega_0 t - \mathbf{k}_t \cdot \mathbf{R}),$$

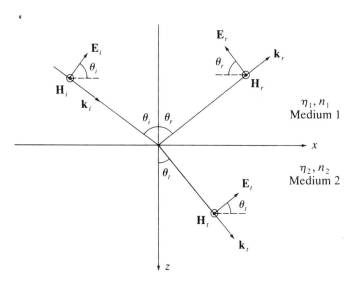

Figure 3.2 Incident, reflected, and transmitted electric fields at the interface between two semiinfinite dielectrics. The polarizations of the electric fields are *parallel* to the plane of incidence (*x-z* plane).

where the directions of $\mathbf{E}_{i0}$, $\mathbf{E}_{r0}$, and $\mathbf{E}_{t0}$ and of $\mathbf{H}_{i0}$, $\mathbf{H}_{r0}$, and $\mathbf{H}_{t0}$ are as shown in the figure. Continuity of the tangential components of $\mathbf{E}$ at the interface requires that at any time and any point,

$$\left(\mathbf{E}_i + \mathbf{E}_r\right)\big|_{\text{along } x} = \mathbf{E}_t\big|_{\text{along } x},$$

which implies

$$E_{i0} \cos \theta_i - E_{r0} \cos \theta_r = E_{t0} \cos \theta_t, \qquad (3.2\text{-}43)$$

where the cosine terms arising from Eqs. (3.2-42) cancel because of the fact that all the $\mathbf{k}$'s are coplanar, that is, $\mathbf{k}_i \cdot \mathbf{R} = \mathbf{k}_r \cdot \mathbf{R} = \mathbf{k}_t \cdot \mathbf{R}$. In Eq. (3.2-43), $E_{i0} = |\mathbf{E}_{i0}|$, $E_{r0} = |\mathbf{E}_{r0}|$, and $E_{t0} = |\mathbf{E}_{t0}|$.

Now, continuity of the tangential components of $\mathbf{H}$ requires that

$$\left(\mathbf{H}_i + \mathbf{H}_r\right)\big|_{\text{along } x} = \mathbf{H}_t\big|_{\text{along } x},$$

which is equivalent to

$$H_{i0} + H_{r0} = H_{t0},$$

or

$$\frac{E_{i0}}{\eta_1} + \frac{E_{r0}}{\eta_1} = \frac{E_{t0}}{\eta_2}$$

or

$$E_{i0} n_1 + E_{r0} n_1 = E_{t0} n_2. \tag{3.2-44}$$

In the preceding set of calculations, $H_{i0} = |\mathbf{H}_{i0}|$, $H_{r0} = |\mathbf{H}_{r0}|$, and $H_{t0} = |\mathbf{H}_{t0}|$; η_1, n_1 and η_2, n_2 denote the intrinsic impedances and refractive indices of media 1 and 2, respectively. From Eqs. (3.2-43) and (3.2-44) we get, after straightforward algebra, the expressions for the *amplitude reflection and transmission coefficients*, $r_\parallel$ and $t_\parallel$, respectively:

$$r_\parallel \triangleq \frac{E_{r0}}{E_{i0}} = \frac{n_2 \cos \theta_i - n_1 \cos \theta_t}{n_2 \cos \theta_i + n_1 \cos \theta_t}, \tag{3.2-45}$$

$$t_\parallel \triangleq \frac{E_{t0}}{E_{i0}} = \frac{2 n_1 \cos \theta_i}{n_2 \cos \theta_i + n_1 \cos \theta_t}. \tag{3.2-46}$$

Perpendicular Polarization

In this case, the $\mathbf{E}$ vectors of all three waves are perpendicular to the plane of incidence, as shown in Figure 3.3. For brevity, we will only state the expressions for the amplitude reflection and transmission coefficients:

$$r_\perp \triangleq \frac{E_{r0}}{E_{i0}} = \frac{n_1 \cos \theta_i - n_2 \cos \theta_t}{n_1 \cos \theta_i + n_2 \cos \theta_t}; \tag{3.2-47}$$

$$t_\perp \triangleq \frac{E_{t0}}{E_{i0}} = \frac{2 n_1 \cos \theta_i}{n_1 \cos \theta_i + n_2 \cos \theta_t}. \tag{3.2-48}$$

In Eqs. (3.2-47) and (3.2-48), E_{i0}, E_{r0}, and E_{t0} denote the amplitudes of the incident, reflected, and transmitted electric fields, respectively, whose directions are shown in Figure 3.3. The derivation of these relations is left as an exercise for the reader.

Equations (3.2-45)–(3.2-48) are called the *Fresnel equations* and dictate plane wave reflection and transmission at the interface between two semi-infinite media characterized by refractive indices n_1 and n_2.

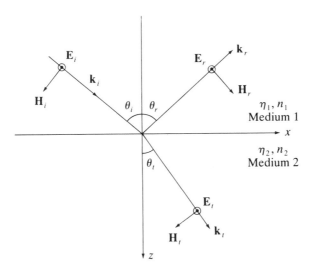

Figure 3.3 Incident, reflected, and transmitted electric fields at the interface between two semiinfinite dielectrics. The polarizations of the electric fields are *perpendicular* to the plane of incidence.

A brief discussion of the Fresnel equations is now in order. Note from Eqs. (3.2-46) and (3.2-48) that the expressions for the transmission coefficients for both parallel and perpendicular polarization are always positive. Referring to Figures 3.2 and 3.3, this means that the transmitted electric field is always in phase with the incident field at the interface. The reflection coefficients [see Eqs. (3.2-45) and (3.2-47)] can, however, be either positive or negative, depending on the values of n_1 and n_2. For instance, if $n_1 > n_2$, $\theta_i < \theta_t$ by Snell's law of refraction [see Eq. (2.2-6b)] and hence $\cos \theta_i > \cos \theta_t$. If $n_1 < n_2$, $\cos \theta_i < \cos \theta_t$. Thus, from Eq. (3.2-47), it follows that, for perpendicular polarization, the reflected wave is in phase with the incident wave at the interface if $n_1 > n_2$, and is π out of phase if $n_1 < n_2$. No such general statement can, however, be made for the reflected wave in the case of parallel polarization.

To illustrate the discussion in the previous paragraph, let us consider plane wave incidence from air at an air–glass interface. Then, taking $n_1 = 1$ and $n_2 = 1.5$, we plot in Figure 3.4 the reflection and transmission coefficients for incident fields with parallel and perpendicular polarization as a function of the incident angle θ_i. The reflection coefficient $r_\perp$ is always negative, because

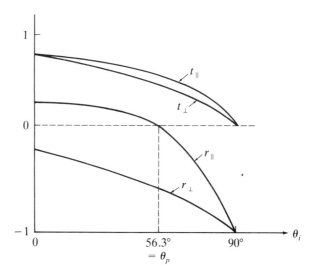

Figure 3.4 Reflection and transmission coefficients for parallel and perpendicular polarization as a function of the angle of incidence θ_i. (θ_p denotes the Brewster angle.)

$n_1 < n_2$. Note, however, that $r_\parallel$ can be positive or negative, depending on the angle of incidence. The incident angle for which $r_\parallel = 0$ is called the *Brewster angle* θ_p. By setting $r_\parallel = 0$ in Eq. (3.2-45), we find

$$n_2 \cos \theta_i = n_1 \cos \theta_t. \qquad (3.2\text{-}49)$$

Using Snell's law [Eq. (2.2-6b)] and Eq. (3.2-49), we get, after straightforward algebra,

$$\sin(\theta_t - \theta_i)\cos(\theta_t + \theta_i) = 0. \qquad (3.2\text{-}50)$$

Because $\theta_t \neq \theta_i$ for $n_2 \neq n_1$, we readily deduce

$$\theta_t + \theta_i = \pi/2, \qquad (3.2\text{-}51)$$

where θ_i and θ_t are related through Snell's law [Eq. (2.2-6b)]. For the example under discussion, $\theta_i = \theta_p = 56.3°$. Unpolarized light, incident at this angle, is reflected as a polarized wave with its **E** vector perpendicular to the plane of incidence.

Linear Wave Propagation

Before ending this subsection, we briefly talk about another value of θ_i that is of physical significance. This angle is the *critical angle* θ_c, which was defined in Eq. (2.2-7) using the ray approach. Recall that for $n_1 > n_2$, any ray incident at an angle greater than the critical angle experiences *total internal reflection*. What is the picture in terms of wave theory? We can answer this question by finding the nature of the transmitted electric field. Note that for $\theta_i > \theta_c$, $\sin \theta_t = (n_1/n_2)\sin \theta_i > 1$, so that

$$\cos \theta_t = \pm j\left[(n_1/n_2)^2 \sin^2 \theta_i - 1\right]^{1/2}.$$

Hence, the transmitted field is of the form

$$\mathbf{E}_t = \text{Re}\left[\mathbf{E}_{t0}\{\exp j\omega_0 t\}\{\exp[-jk_t(x \sin \theta_t + z \cos \theta_t)]\}\right]$$

$$= \text{Re}\left[\mathbf{E}_{t0}\{\exp j\omega_0 t\}\left\{\exp\left[-jk_t\frac{n_1}{n_2}x \sin \theta_i\right]\right\}\right.$$

$$\left.\times\left\{\exp\left(-k_t z\left[\left(\frac{n_1}{n_2}\right)^2 \sin^2 \theta_i - 1\right]^{1/2}\right)\right\}\right], \qquad (3.2\text{-}52)$$

where we have only retained the real exponential in z with a negative argument to prevent nonphysical solutions. As is easily seen from Eq. (3.2-52), this represents a field propagating in the x direction, with an exponentially decreasing amplitude in the z direction. Such a wave is called a *surface wave* along x and an *evanescent wave* in the z direction (see Figure 3.5).

At the same time, note that the reflection and transmission coefficients [see Eqs. (3.2-45)–(3.2-48)], become, in general, complex. The phase of the reflected wave at the interface is now a function of the angle of incidence. If we consider a collection of plane waves, travelling in different directions, to be incident on the interface at angles larger than the critical angle, each plane wave experiences total internal reflection, and the reflection coefficient for each is different in phase from that of the others. Now, we show in the next section that a beam of light can be visualized as a collection of plane waves of weighted amplitudes, called the *angular*

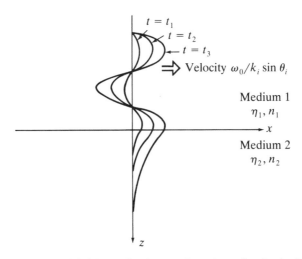

Figure 3.5 Variation of field amplitude as a function of z for incidence at an angle greater than the critical angle. In medium 1, the incident and reflected fields form a standing wave pattern in the z direction. In medium 2, there is an evanescent wave. The entire pattern travels to the right with a phase velocity $\omega_0/(k_i \sin \theta_i)$.

plane-wave spectrum, travelling in different directions. Thus, for a beam incident at the interface, we can reconstruct the reflected beam by carefully adding the complex amplitudes of the reflected plane waves corresponding to the incident plane wave spectrum at the interface. The net result is a reflected beam that is laterally shifted upon reflection, as depicted in Figure 3.6. The incident beam appears to have been reflected from a fictitious plane displaced from the interface. This effect is called the *Goos–Hänchen shift*, after the names of its discoverers. A detailed analysis of the Goos–Hänchen shift is outside the scope of this book. We refer interested readers to Lee (1986).

Before ending this subsection, we will discuss (using an example) the effect of multiple-beam interference when a plane wave is incident on a plane-parallel plate, using the knowledge of the reflection and transmission coefficients developed earlier. This arrangement is conventionally called the Fabry–Perot etalon and forms an integral part of many optical systems, including the laser cavity, to be discussed in more detail in Chapter 7.

Linear Wave Propagation

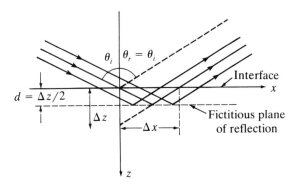

Figure 3.6 Incident and reflected *beams* when the incident angle exceeds the critical angle, showing the lateral displacement (Goos–Hänchen shift) Δx and the location of the fictitious reflecting plane. Note that $\Delta x = \Delta z \tan \theta_i$.

Example 3.1 Multiple-Beam Interference in a Plane-Parallel Plate: The Fabry–Perot Etalon

Consider a plane-parallel plate of thickness l, having refractive index n_2 as shown in Figure 3.7. Let E_i be the complex electric field amplitude of the incident wave, assumed to be linearly polarized, with the polarization vector either parallel or perpendicular to the plane of incidence. Then the successive reflected rays from the *Fabry-Perot etalon* can be expressed by the phasors

$$E_1 = rE_i,$$

$$E_2 = tr't'E_ie^{-j\delta} = tt'r'E_ie^{-j\delta},$$

$$E_3 = tr'^3t'E_ie^{-j2\delta} = tt'r'^3E_ie^{-j2\delta}, \qquad (3.2\text{-}53)$$

$$\vdots$$

$$E_N = tt'r'^{(2N-3)}E_ie^{-j(N-1)\delta},$$

where r and t are the reflection and transmission coefficients for *rays* incident from n_1 toward n_2, and r' and t' are the correspond-

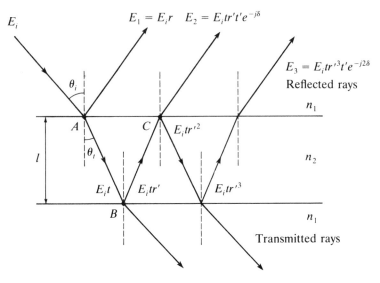

Figure 3.7 Multiple-beam interference of an incident wave of amplitude E_i from a plane-parallel plate.

ing quantities for rays travelling from n_2 toward n_1; δ is the phase difference between adjacent reflected rays, given by

$$\delta = \frac{4\pi}{\lambda_v} n_2 l \cos \theta_t, \qquad (3.2\text{-}54)$$

where λ_v is the incident wavelength in vacuum and θ_t is the transmission angle. [For notational convenience, we have eliminated the subscript p (which denotes phasors) after the Es in this discussion.] We can then write the expression for the resultant reflected amplitude E_r as

$$E_r = \sum_{1}^{N} E_j$$

$$= E_i \Big[r + tt'r'e^{-j\delta} \Big[1 + (r'^2 e^{-j\delta}) + (r'^2 e^{-j\delta})^2 + \cdots$$

$$+ (r'^2 e^{-j\delta})^{N-2} \Big] \Big]. \quad (3.2\text{-}55)$$

If $|r'^2 e^{-j\delta}| < 1$, and if $N \to \infty$ (i.e., the number of reflected rays is large), it is possible to reexpress Eq. (3.2-55) in closed form as

$$E_r = E_i \left[r + \frac{r' t t' e^{-j\delta}}{1 - r'^2 e^{-j\delta}} \right]. \tag{3.2-56}$$

Now, making use of the fact that

$$r = -r' \quad \text{[see Eqs. (3.2-45) and (3.2-47)]}, \tag{3.2-57a}$$

$$tt' + r^2 = 1 \quad \text{[see Eqs. (3.2-45)–(3.2-48)]}, \tag{3.2-57b}$$

we can simplify Eq. (3.2-56) to

$$E_r = E_i \left[\frac{r(1 - e^{-j\delta})}{1 - r^2 e^{-j\delta}} \right]. \tag{3.2-58}$$

The fraction of the incident intensity that is reflected is

$$\frac{I_r}{I_i} = \frac{|E_r|^2}{|E_i|^2} = \frac{F \sin^2(\delta/2)}{1 + F \sin^2(\delta/2)}, \tag{3.2-59}$$

where $F = [2r/(1 - r^2)]^2$ is known as the *coefficient of finesse*.

A similar treatment of the transmitted rays leads to the following expression for the fraction of the transmitted intensity:

$$\frac{I_t}{I_i} = \frac{1}{1 + F \sin^2(\delta/2)}. \tag{3.2-60}$$

Note that from Eqs. (3.2-59) and (3.2-60), one obtains $I_r + I_t = I_i$, which is the well-known conservation of energy.

The term $[1 + F \sin^2(\delta/2)]^{-1}$ is known as the *Airy function* $A(\delta)$. It represents the characteristics of I_t/I_i as a function of the phase difference δ and is plotted in Figure 3.8 for various values of F and therefore, r. Note that the transmitted intensity peaks around the points where $\delta/2 = m\pi$ (where m is an integer), that is,

$$2n_2 l \cos \theta_t = m\lambda_v. \tag{3.2-61}$$

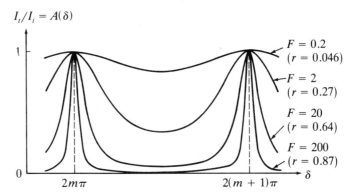

Figure 3.8 Plot of transmission versus phase difference in multiple-beam interferometer.

Also, as r, and hence F, increases, the width of the transmission peaks reduces.

The Fabry–Perot etalon can be used in precision wavelength measurements when two different frequencies are present in the incident light. This is achieved by the arrangement shown in Figure 3.9, where the etalon is illuminated by a collimated beam from an extended source. The transmitted waves are focussed by a lens on a screen, where they interfere to form a fringe pattern. For light of a specific wavelength λ_v, recall that the incident light is completely transmitted when Eq. (3.2-61) is satisfied.

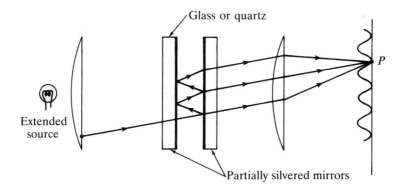

Figure 3.9 Fabry–Perot interferometer.

For large values of F, when θ_t is slightly different from the values given by Eq. (3.2-61), the transmittivity is very small. For a given wavelength, the fringe pattern will correspond to different values of m. The sharpness of the fringes will depend on the value of F. It is easy to see that with a high F, two illuminating wavelengths will give easily discernible fringes with different periodicities. This is the way two very closely spaced wavelengths (as in hyperfine splitting) can be resolved.

3.2.4 Dispersion

Thus far, we have studied the propagation of waves using the wave equation [e.g., Eqs. (3.2-10), (3.2-11), and (3.2-13)] whose properties, described by μ, ϵ, or v, have been implicitly assumed to be a constant. In many cases, however, the material properties (e.g., ϵ or, equivalently, n) may be a function of the frequency ω of the wave propagating through the medium. This dependence of n on ω (or, correspondingly, the wavelength λ of the wave), which is called *dispersion*, is governed by the interplay of various electric polarization mechanisms contributing at the particular frequency. The following are examples of such mechanisms:

(a) *orientational polarization, where the dipoles align themselves when an electric field is externally applied;*
(b) *electronic polarization, where the applied field distorts the electron cloud with respect to the nucleus;*
(c) *ionic or atomic polarization, where the positive and negative ions undergo a relative shift with the application of an external field.*

By employing a simple classical model, one can understand the nature of dispersion arising from electronic polarization. We will not discuss this here, but refer readers to Hecht and Zajac (1975) for a detailed treatment. Rather, in what follows, we will outline a method to analyze wave propagation in dispersive media once the *dispersion relation* is available. By the term dispersion relation, we will mean the dependence of the angular frequency ω on the propagation constant k of the wave. Dispersion relations are available in various forms, for example, the dependence of the refractive index n on the wavelength λ or the frequency ω. In all cases, these variations can be translated to the dependence of ω on k, describable in general as

$$\omega = W(k). \qquad (3.2\text{-}62)$$

As discussed in Section 3.2.1, a dispersion relation corresponding to the wave equation Eq. (3.2-13) is of the form

$$\omega = vk. \qquad (3.2\text{-}63)$$

If we plot ω as a function of k, the graph is a straight line. As mentioned earlier, the medium of propagation is nondispersive. On the other hand, if the dispersion curve is not linear, we have a *dispersive* medium. Unlike the nondispersive case, the slope of the dispersion curve is no longer a constant, but depends on k, and hence ω. We may redefine dispersive and nondispersive waves as follows:

$$\frac{d\omega}{dk} \neq \frac{\omega}{k} \quad \text{(dispersive waves)}:$$

$$\frac{d\omega}{dk} = \frac{\omega}{k} = \text{constant} \quad \text{(nondispersive waves)}.$$

Example 3.2
Suppose that the dispersion relation is

$$\omega = vk - \gamma k^3 \qquad (3.2\text{-}64)$$

(see Figure 3.10). Here, $\omega/k = v - \gamma k^2$ and $d\omega/dk = v - 3\gamma k^2$; hence, the medium is dispersive.

Recall from elementary electromagnetics that ω/k is called the *phase velocity* of the wave, as it corresponds to the velocity of the *phase fronts*, or wavefronts (see Section 3.2.1). On the other

Linear Wave Propagation

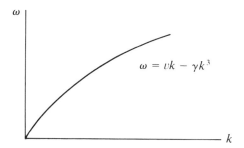

Figure 3.10 Dispersion curve ω versus for dispersion relation $\omega = vk - \gamma k^3$.

hand, $d\omega/dk$ is called the *group velocity*. The name comes from the fact that if two slightly different frequencies are propagating through a medium, we can visualize this as the propagation of an envelope on a carrier. To see this very simply, consider a wavefunction $\psi(z, t)$ of the form

$$\psi(z, t) = A\big[\sin(\omega_0 t - k_0 z) + \sin((\omega_0 + \Delta\omega)t - (k_0 + \Delta k)z)\big]$$

$$= 2A \sin(\omega_0 t - k_0 z)\cos((\Delta\omega/2)t - (\Delta k/2)z). \quad (3.2\text{-}65)$$

If $\Delta\omega \ll \omega_0$ and $\Delta k \ll k_0$, we can think of ψ as comprising an envelope $\cos((\Delta\omega/2)t - (\Delta k/2)z)$ riding on a carrier $\sin(\omega_0 t - k_0 z)$. The velocity of the envelope, or *group*, is given by $\Delta\omega/\Delta k$ (which in the limit becomes $d\omega/dk$) and is the group velocity. Thus, in a dispersive medium, the group velocity is not equal to the phase velocity.

Given the dispersion relation Eq. (3.2-62), how do we analyze the propagation of waves through the medium? The answer lies in first finding the corresponding partial differential equation (PDE) that can describe the phenomenon. To do this, we will take recourse to a simple procedure we followed before in order to find the dispersion relation Eq. (3.2-15) [or Eq. (3.2-63)] from the wave equation. Thinking of the most elementary form of a travelling wave [e.g., $\exp(j(\omega t - kz))$], we can find the dispersion relation from the wave equation by noting that the operations of differentiation with respect to time and space amount to multiplying the exponential by $j\omega$ and $-jk$, respectively. Thus, in order to derive the PDE, we simply replace

$$\omega \to -j\frac{\partial}{\partial t}, \qquad k \to j\frac{\partial}{\partial z} \qquad (3.2\text{-}66)$$

to find the operator equation. This operating on a wavefunction $\psi(z, t)$ gives the required PDE.

Example 3.3
The PDE corresponding to the dispersion relation Eq. (3.2-64) is

$$\frac{\partial \psi}{\partial t} + v\frac{\partial \psi}{\partial z} + \gamma\frac{\partial^2 \psi}{\partial z^3} = 0. \qquad (3.2\text{-}67)$$

We quote here a result without proof: for real wavefunctions, the underlying dispersion relation Eq. (3.2-62) must be an odd function of k. For details, we refer readers to Korpel and Banerjee (1984). Another example of finding the PDE from the dispersion relation can be found in Chapter 8 in connection with the propagation of (complex) envelopes in a dispersive medium described by a generalized dispersion relation. It will turn out that in this case, the dispersion relation need not be an odd function, because the wavefunction, which in this case is the envelope, may be complex.

3.2.5 Eikonal Equations
So far, we have only considered the propagation of electromagnetic waves having a constant amplitude. An alternate treatment for this case is provided by the eikonal (or ray) equations in optics, which has its roots in the ray approach, where light is treated in terms of the corpuscular, or particle, theory. In this

section, we derive the equations governing ray propagation by starting from the wave equation. We will see that the wave approach, which explains diffraction of light, yields the modified eikonal equations, which take diffraction into account. A further motivation for deriving these equations is that they form an alternate representation of the Fresnel diffraction formula, which we derive in the next section.

Consider, therefore, the wave equation in three dimensions and in Cartesian coordinates in the form

$$\frac{1}{v^2}\frac{\partial^2 \psi}{\partial t^2} - \frac{\partial^2 \psi}{\partial x^2} - \frac{\partial^2 \psi}{\partial y^2} - \frac{\partial^2 \psi}{\partial z^2} = 0 \qquad (3.2\text{-}68)$$

and assume ψ to be time-harmonic, that is,

$$\psi(x, y, z, t) = \text{Re}\{\psi_p(x, y, z)e^{j\omega_0 t}\}. \qquad (3.2\text{-}69)$$

Substituting Eq. (3.2-69) into Eq. (3.2-68), we get the *Helmholtz equation* for ψ_p,

$$\frac{\partial^2 \psi_p}{\partial x^2} + \frac{\partial^2 \psi_p}{\partial y^2} + \frac{\partial^2 \psi_p}{\partial z^2} + k_0^2 \psi_p = 0, \qquad k_0 = \frac{\omega_0}{v}. \quad (3.2\text{-}70)$$

Now, putting

$$\psi_p(x, y, z) = a(x, y, z)\exp[-j\tilde{\phi}(x, y, z)] \qquad (3.2\text{-}71)$$

into Eq. (3.2-70) and separating the real and imaginary parts, we derive the coupled sets of equations

$$a\left[\left(\frac{\partial \tilde{\phi}}{\partial x}\right)^2 + \left(\frac{\partial \tilde{\phi}}{\partial y}\right)^2 + \left(\frac{\partial \tilde{\phi}}{\partial z}\right)^2\right] = k_0^2 a + \left[\frac{\partial^2 a}{\partial x^2} + \frac{\partial^2 a}{\partial y^2} + \frac{\partial^2 a}{\partial z^2}\right]$$

$$(3.2\text{-}72)$$

Physical Optics

and

$$2\left[\left(\frac{\partial a}{\partial x}\right)\left(\frac{\partial \tilde{\phi}}{\partial x}\right) + \left(\frac{\partial a}{\partial y}\right)\left(\frac{\partial \tilde{\phi}}{\partial y}\right) + \left(\frac{\partial a}{\partial z}\right)\left(\frac{\partial \tilde{\phi}}{\partial z}\right)\right]$$

$$+ a\left[\frac{\partial^2 \tilde{\phi}}{\partial x^2} + \frac{\partial^2 \tilde{\phi}}{\partial y^2} + \frac{\partial^2 \tilde{\phi}}{\partial z^2}\right] = 0. \tag{3.2-73}$$

We can rewrite Eq. (3.2-72) as

$$(\nabla\tilde{\phi}) \cdot (\nabla\tilde{\phi}) = k_0^2 + \frac{1}{a}\nabla^2 a. \tag{3.2-74}$$

Note that the gradient of $\tilde{\phi}$ ($= \nabla\tilde{\phi}$) gives the direction of the maximum space rate of change of $\tilde{\phi}$, that is, the local direction of *rays*. Without the last term on the RHS, Eq. (3.2-74) simply states that the sum of the squares of the projections of $\nabla\tilde{\phi}$ on the $x, y,$ and z axes, namely, k_x, k_y, and k_z, is equal to k_0^2, as expected. The last term denotes the contribution from *diffraction*, which causes *bending of light*, and is absent if we use the ray approach. The relevance of this term will become clearer when we derive the Fresnel diffraction formula in the following section.

Now, by multiplying Eq. (3.2-73) by a, we can restate the equation as

$$\nabla \cdot \left(a^2 \nabla\tilde{\phi}\right) = 0. \tag{3.2-75}$$

To understand the physical significance of Eq. (3.2-75), we will integrate both sides over a control volume and invoke the divergence theorem, Eq. (3.1-5), to convert the volume integral to an integral over the surface enclosing the control volume. The integrand $a^2 \nabla\tilde{\phi}$ is indicative of the directed flux flowing out, and the integral is equal to zero. This makes sense because there are no sources within the control volume. Equation (3.2-75) can be directly rederived using the ray approach.

Equations (3.2-74) and (3.2-75) constitute the *eikonal equations* in optics and in the presence of diffraction.

3.3 Derivation and Use of the Fresnel Diffraction Formula

In the previous section, we introduced diffraction as the bending of light. The name was coined in the seventeenth century by Grimaldi, who studied the deviation of light from rectilinear propagation. This is characteristically a wave phenomenon, occurring whenever a portion of the wavefront is obstructed in some way or, more generally, whenever there is amplitude distribution along the wavefront. For instance, when light encounters an aperture, the light spreads or diffracts as it travels, and the amount of spread increases with propagation distance.

For a heuristic derivation of the amount of spread, consider light emanating from an aperture of width d, as shown in Figure 3.11. Lines AA' and BB' represent rays emanating from the endpoints of the aperture. Quantum mechanics relates the minimum uncertainty in position Δx of a quantum to the uncertainty in its momentum Δp_x according to

$$\Delta x \, \Delta p_x \sim \hbar, \qquad (3.3\text{-}1)$$

where $\hbar = h/2\pi$ [$h = 6.625 \times 10^{-34}$ J-s (Planck's constant)]. Now, in our problem $\Delta x = d$, because the *quantum* of light can emerge from any point on the aperture. Hence, by Eq. (3.3-1),

$$\Delta p_x \sim \frac{\hbar}{d}. \qquad (3.3\text{-}2)$$

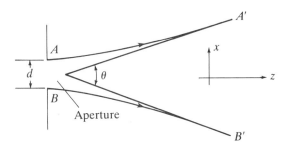

Figure 3.11 Geometry for heuristic determination of the angle of spread θ during diffraction.

We define the angle θ of spread, assumed small, as

$$\theta \sim \frac{\Delta p_x}{p_z} \sim \frac{\Delta p_x}{p_0}, \qquad (3.3\text{-}3)$$

where p_z and p_0 represent the z component of the momentum, and the momentum of the quantum, respectively. But $p_0 = \hbar k_0$, where k_0 is the propagation constant; hence,

$$\theta \sim \frac{1}{k_0 d} = \frac{1}{2\pi} \frac{\lambda_0}{d}, \qquad (3.3\text{-}4)$$

where λ_0 denotes the wavelength in the medium of propagation of the light. Thus, the angle of spread is inversely proportional to the aperture width (in number of wavelengths). As we shall see time and time again, the parameter λ_0/d will feature in many problems of diffraction.

3.3.1 The Spatial Transfer Function and Fresnel Diffraction

The preceding simple argument gives us a heuristic picture of the effect of diffraction. However, it does not tell us the exact nature of the amplitude and/or phase distribution during propagation. To find this, we can either employ the eikonal equations derived in the previous section to find the amplitude and phase distributions individually, or else find a way to predict the complex amplitude all at once. In this section, we try the latter approach, starting from the wave equation, Eq. (3.2-68). We assume that the total wavefunction $\psi(x, y, z, t)$ comprises a *complex envelope* $\psi_e(x, y, z)$ riding on a carrier of frequency ω_0 and propagation constant k_0 that propagates in the $+z$ direction:

$$\psi(x, y, z; t) = \text{Re}\{\psi_e(x, y, z)\exp[\,j(\omega_0 t - k_0 z)]\}, \qquad \frac{\omega_0}{k_0} = v.$$

$$(3.3\text{-}5)$$

Recall that the complex envelope ψ_e is related to the phasor ψ_p in

Eq. (3.2-69) as

$$\psi_p(x, y, z) = \psi_e(x, y, z)\exp(-jk_0z), \qquad (3.3\text{-}6)$$

and we will use one or the other according to convenience. Note also that

$$|\psi_e(x, y, z)| = |\psi_p(x, y, z)| = a(x, y, z), \qquad (3.3\text{-}7)$$

using Eq. (3.2-71), so that by writing

$$\psi_e(x, y, z) = a(x, y, z)\exp[-j\phi(x, y, z)], \qquad (3.3\text{-}8)$$

it readily follows that

$$\tilde{\phi}(x, y, z) = \phi(x, y, z) + k_0z. \qquad (3.3\text{-}9)$$

Substituting Eq. (3.3-6) into Eq. (3.2-70) and assuming that $\psi_e(x, y, z)$ is a *slowly varying* function of z (the direction of propagation) in the sense that

$$\frac{|\partial^2\psi_e/\partial z^2|}{|\partial\psi_e/\partial z|} \ll k_0, \qquad (3.3\text{-}10)$$

we obtain the *paraxial wave equation*,

$$2jk_0\frac{\partial\psi_e}{\partial z} = \frac{\partial^2\psi_e}{\partial x^2} + \frac{\partial^2\psi_e}{\partial y^2}. \qquad (3.3\text{-}11)$$

Equation (3.3-11) describes the propagation of the envelope $\psi_e(x, y, z)$, starting from an initial profile

$$\psi_e|_{z=0} = \psi_{e0}(x, y). \qquad (3.3\text{-}12)$$

In what follows, we solve Eq. (3.3-11) using Fourier transform techniques. As we will see shortly, this helps us understand diffraction from a simple transfer function concept of wave propagation, and also yields the Fresnel diffraction formula.

Assuming ψ_e to be Fourier-transformable, we employ the definition of the Fourier transform enunciated in Chapter 1 [see

Eq. (1.1-5)] and its properties (see Table 1.2) to Fourier-transform Eq. (3.3-11). This gives

$$\frac{d\Psi_e}{dz} = \frac{j}{2k_0}\left(k_x^2 + k_y^2\right)\Psi_e, \qquad (3.3\text{-}13)$$

where $\Psi_e(k_x, k_y; z)$ is the Fourier transform of $\psi_e(x, y, z)$. We now readily solve Eq. (3.3-13) to get

$$\Psi_e(k_x, k_y; z) = \Psi_{e0}(k_x, k_y)\exp\left[\frac{j\left(k_x^2 + k_y^2\right)z}{2k_0}\right], \qquad (3.3\text{-}14)$$

where

$$\Psi_{e0}(k_x, k_y) = \Psi_e(k_x, k_y; z)\big|_{z=0} = \mathscr{F}_{xy}\{\psi_{e0}(x, y)\}. \qquad (3.3\text{-}15)$$

We remark that $\Psi_p(k_x, k_y; z) = \Psi_e(k_x, k_y; z)e^{-jk_0z}$ is sometimes referred to as the *angular plane-wave spectrum* of $\psi_e(x, y, z)$ [Goodman (1968)]. We can interpret Eq. (3.3-14) in the following way: Consider a linear system with $\Psi_{e0}(k_x, k_y)$ as its input spectrum (i.e., at $z = 0$) and where the output spectrum is $\Psi_e(k_x, k_y; z)$. Then, the *spatial frequency response* of the linear system is given by

$$\frac{\Psi_e}{\Psi_{e0}} \triangleq H(k_x, k_y; z) = \exp\left[\frac{j\left(k_x^2 + k_y^2\right)z}{2k_0}\right]. \qquad (3.3\text{-}16)$$

We will call $H(k_x, k_y; z)$ the *spatial transfer function of propagation* of light through a distance z in the medium as shown in Figure 3.12. The *spatial impulse response* is given by

$$h(x, y, z) = \mathscr{F}_{xy}^{-1}\{H(k_x, k_y; z)\}$$

$$= \frac{jk_0}{2\pi z}\exp\left[-\frac{j(x^2 + y^2)k_0}{2z}\right]. \qquad (3.3\text{-}17)$$

Thus, by taking the inverse Fourier transform of Eq. (3.3-14) and

Fresnel Diffraction Formula

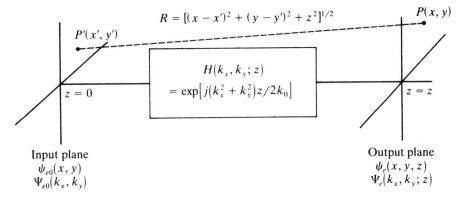

Figure 3.12 Block-diagrammatic representation of the spatial transfer function of propagation. The input and output planes have primed and unprimed coordinate systems, respectively.

using Eq. (3.3-17), we obtain

$$\psi_e(x, y, z) = \psi_{e0}(x, y) * h(x, y, z)$$

$$= \frac{jk_0}{2\pi z} \int_{-\infty}^{\infty}\int_{-\infty}^{\infty} \psi_{e0}(x', y')$$

$$\times \exp\left[-j\frac{k_0}{2z}\{(x - x')^2 + (y - y')^2\}\right] dx'\, dy', \quad (3.3\text{-}18)$$

where $*$ denotes convolution. The way Eq. (3.3-18) has been written indicates that the transverse coordinates are x', y' in the input plane and x, y in the output plane at a distance z away from the input plane (see Figure 3.12). Using Eq. (3.3-6), we can write, from Eq. (3.3-18),

$$\psi_p(x, y, z) = \frac{jk_0}{2\pi z} \exp[-jk_0 z] \int_{-\infty}^{\infty}\int_{-\infty}^{\infty} \psi_{p0}(x', y')$$

$$\times \exp\left[-j\frac{k_0}{2z}\{(x - x')^2 + (y - y')^2\}\right] dx'\, dy', \quad (3.3\text{-}19a)$$

where

$$\psi_{p0}(x, y) = \psi_p(x, y, z)\big|_{z=0} = \psi_e(x, y, z)\big|_{z=0} = \psi_{e0}(x, y). \quad (3.3\text{-}19b)$$

Equation (3.3-18) or (3.3-19) is termed the *Fresnel diffraction formula* and describes the Fresnel diffraction of a beam during propagation and having an arbitrary initial complex amplitude profile ψ_{e0}.

3.3.2 Illustrative Examples

Example 3.4 Point Source

$$\psi_{e0}(x, y) = \delta(x)\delta(y). \qquad (3.3-20)$$

By Eq. (3.3-18),

$$\psi_e(x, y, z) = [\delta(x)\delta(y)] * h(x, y, z)$$

$$= \frac{jk_0}{2\pi z} \exp\left[-\frac{j(x^2 + y^2)k_0}{2z}\right]. \qquad (3.3-21)$$

Hence, using Eq. (3.3-6),

$$\psi_p(x, y, z) = \frac{jk_0}{2\pi z} \exp\left[-jk_0 z - \frac{j(x^2 + y^2)k_0}{2z}\right]. \qquad (3.3-22)$$

Consider the argument of the exponent in Eq. (3.3-22). Note that it is proportional to $z + (x^2 + y^2)/2z$. Recall that, using the binomial theorem,

$$(z^2 + x^2 + y^2)^{1/2} = z\left[1 + \frac{x^2 + y^2}{z^2}\right]^{1/2} \simeq z + \frac{x^2 + y^2}{2z}, \qquad (3.3-23)$$

provided

$$x^2 + y^2 \ll z^2. \qquad (3.3-24)$$

The condition just stated is called the *paraxial approximation* in optics and was introduced in Chapter 2. We will use it very often throughout the book, during our analysis of optical propagation. It means that we only consider small excursions around the axis of propagation (z). In actuality, this is often not the case and thus, strictly speaking, we need to include higher-order terms in the expansion, Eq. (3.3-23). These higher-order terms lead to *aberrations*. The other part of the paraxial approximation, commensurate with Eq. (3.3-10), can be stated as

$$k_x^2 + k_y^2 \ll k_0^2, \tag{3.3-25}$$

meaning that the x and y components of the propagation vector of a wave are relatively small, because the wave is assumed to propagate *nominally* in the z direction.

Incorporating Eq. (3.3-23) into Eq. (3.3-22), we have

$$\psi_p(x, y, z) \simeq \frac{jk_0}{2\pi z} \exp\left(-jk_0[z^2 + x^2 + y^2]^{1/2}\right)$$

$$\simeq \frac{jk_0}{2\pi R} \exp(-jk_0 R); \tag{3.3-26}$$

thus Eq. (3.3-21) represents the paraxial approximation to a spherical wave, as expected.

Example 3.5 Plane Wave

$$\psi_{e0}(x, y) = 1. \tag{3.3-27}$$

Then

$$\Psi_{e0}(k_x, k_y) = 4\pi^2 \delta(k_x)\delta(k_y),$$

so that, using Eq. (3.3-14),

$$\Psi_e(k_x, k_y; z) = 4\pi^2 \delta(k_x)\delta(k_y)\exp\left[\frac{j\left(k_x^2 + k_y^2\right)z}{2k_0}\right]$$

$$= 4\pi^2 \delta(k_x)\delta(k_y)\exp\left[\frac{j\left(k_x^2 + k_y^2\right)z}{2k_0}\right]\Bigg|_{k_x=0=k_y}$$

$$= 4\pi^2 \delta(k_x)\delta(k_y)$$

$$= \Psi_{e0}. \tag{3.3-28}$$

Physical Optics

Hence,

$$\psi_e = \psi_{e0} = 1,$$

that is, the plane wave travels undiffracted, as expected.

Example 3.6 Gaussian Beam

$$\psi_{e0}(x, y) = \exp\left[-\frac{(x^2 + y^2)}{w_0^2}\right].$$ (3.3-29)

Here,

$$\Psi_{e0}(k_x, k_y) = \pi w_0^2 \exp\left[-\frac{(k_x^2 + k_y^2)w_0^2}{4}\right],$$

hence,

$$\Psi_e(k_x, k_y; z) = \Psi_{e0}(k_x, k_y)H(k_x, k_y; z)$$

$$= \pi w_0^2 \exp\left[-\frac{(k_x^2 + k_y^2)w_0^2}{4}\right]\exp\left[\frac{j(k_x^2 + k_y^2)z}{2k_0}\right]$$

$$= \pi w_0^2 \exp\left[\frac{j(k_x^2 + k_y^2)q}{2k_0}\right],$$ (3.3-30)

where

$$q \triangleq z + \frac{jk_0 w_0^2}{2} \triangleq z + jz_R$$ (3.3-31)

is called the *q parameter* of the Gaussian beam. The propagation of a Gaussian beam is completely described by the transformation of its q parameter. For instance, it is easy to see from Eqs. (3.3-30) and (3.3-31) that if a Gaussian beam propagates through a distance

d in free space, its q changes to q', where

$$q' = q + d. \tag{3.3-32}$$

Now, to find the evolution of the Gaussian beam during propagation, we need to find the inverse Fourier transform of Eq. (3.3-30). Note that this equation is of the same form as Eq. (3.3-16), with z replaced by q. Hence, using Eq. (3.3-17),

$$\psi_e(x, y, z) = \frac{jk_0 w_0^2}{2q} \exp\left[-\frac{j(x^2 + y^2)k_0}{2q}\right]$$

$$= \left(\frac{w_0}{w(z)}\right) \exp\left\{-\frac{x^2 + y^2}{w^2(z)}\right\}$$

$$\times \exp\left\{-\frac{jk_0}{2R}(x^2 + y^2)\right\} \exp(-j\phi), \tag{3.3-33}$$

where

$$w^2(z) = \frac{2z_R}{k_0}\left[1 + \left(\frac{z}{z_R}\right)^2\right], \tag{3.3-34}$$

$$R(z) = \frac{z^2 + z_R^2}{z}, \tag{3.3-35}$$

$$\phi(z) = -\tan^{-1}\left(\frac{z}{z_R}\right). \tag{3.3-36}$$

Inspection of Eqs. (3.3-33)–(3.3-36) shows that the magnitude of the diffracted Gaussian profile is still a Gaussian, albeit with a decrease in amplitude during propagation and an increase in its width w (initially equal to the *waist size* w_0). Observe that at $z = z_R = k_0 w_0^2/2$, $w^2 = 2w_0^2$; this distance is called the *Rayleigh length*. The remaining terms indicate *phase curvature*, with the radius of curvature R depending on the distance of propagation and an additional phase shift ϕ. Note that the radius of curvature is always positive and indicates diverging wavefronts. Furthermore, by differentiating Eq. (3.3-35) with respect to z, we can find the position of the minimum radius of curvature to be at $z = z_R$. Thus, starting at $z = 0$, the radius of curvature first decreases from infinity (plane wavefronts) to a minimum value before starting to increase again. For large values of z, the radius of curvature is approximately equal to z, corresponding to spherical wavefronts, because from these distances, the original profile (at $z = 0$) appears to be a point source. Correspondingly, from Eqs. (3.3-33) and (3.3-34), the amplitude decreases according to $1/z$, as expected.

We comment, in passing, that the preceding analysis could also be performed by starting from the eikonal equations [Eqs. (3.2-74) and (3.2-75)]. The computations are, in general, more complicated; however, a priori assumptions of the nature of the amplitude and phase profiles make the problem more tractable. It can be checked that the last term on the RHS of Eq. (3.2-74) indeed represents the effects of diffraction—without it, the Gaussian beam propagates undiffracted, as predicted from ray optics.

Example 3.7 Plane Wave through a Rectangular Aperture

$$\psi_{e0}(x, y) = \text{rect}\left(\frac{x}{l}\right)\text{rect}\left(\frac{y}{l}\right). \tag{3.3-37}$$

Using Eq. (3.3-18),

$$\psi_e(x, y, z) = \frac{jk_0}{2\pi z} \int_{-\infty}^{\infty}\int_{-\infty}^{\infty} \text{rect}\left(\frac{x'}{l}\right)\text{rect}\left(\frac{y'}{l}\right)$$

$$\times \exp\left[-j\frac{k_0}{2z}\{(x - x')^2 + (y - y')^2\}\right] dx'\,dy'$$

$$= \frac{jk_0}{2\pi z}g(x)g(y), \tag{3.3-38}$$

where

$$g(x) = \int_{-l/2}^{l/2} \exp\left[-j\frac{k_0}{2z}(x'-x)^2\right] dx',$$

$$(3.3\text{-}39)$$

$$g(y) = \int_{-l/2}^{l/2} \exp\left[-j\frac{k_0}{2z}(y'-y)^2\right] dy'.$$

These integrals are substantially simplified by the change of variables

$$\zeta = \left(\frac{k_0}{\pi z}\right)^{1/2}(x'-x), \qquad \eta = \left(\frac{k_0}{\pi z}\right)^{1/2}(y'-y), \quad (3.3\text{-}40a)$$

yielding

$$g(x) = \left\{\frac{\pi z}{k_0}\right\}^{1/2} \int_{\zeta_1}^{\zeta_2} \exp\left[-j\frac{\pi}{2}\zeta^2\right] d\zeta,$$

$$(3.3\text{-}40b)$$

$$g(y) = \left\{\frac{\pi z}{k_0}\right\}^{1/2} \int_{\eta_1}^{\eta_2} \exp\left[-j\frac{\pi}{2}\eta^2\right] d\eta,$$

where

$$\zeta_1 = -\left(\frac{k_0}{\pi z}\right)^{1/2}\left(\frac{l}{2}+x\right), \qquad \eta_1 = -\left(\frac{k_0}{\pi z}\right)^{1/2}\left(\frac{l}{2}+y\right),$$

$$(3.3\text{-}40c)$$

$$\zeta_2 = \left(\frac{k_0}{\pi z}\right)^{1/2}\left(\frac{l}{2}-x\right), \qquad \eta_2 = \left(\frac{k_0}{\pi z}\right)^{1/2}\left(\frac{l}{2}-y\right).$$

We can recast Eqs. (3.3-40b) in the forms

$$g(x) = \left(\frac{\pi z}{k_0}\right)^{1/2}[\{C(\zeta_2) - C(\zeta_1)\} - \{jS(\zeta_2) - S(\zeta_1)\}],$$

$$(3.3\text{-}41)$$

$$g(y) = \left(\frac{\pi z}{k_0}\right)^{1/2}[\{C(\eta_2) - C(\eta_1)\} - j\{S(\eta_2) - S(\eta_1)\}],$$

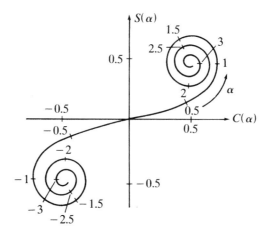

Figure 3.13 The Cornu spiral, used to compute the Fresnel diffraction pattern of a rectangular aperture.

where $C(\alpha)$ and $S(\alpha)$ denote the *Fresnel integrals*, defined by

$$C(\alpha) = \int_0^\alpha \cos\frac{\pi t^2}{2}\, dt, \qquad S(\alpha) = \int_0^\alpha \sin\frac{\pi t^2}{2}\, dt. \quad (3.3\text{-}42)$$

We can evaluate the Fresnel integrals by using the *Cornu spiral* (see Figure 3.13), which is a simultaneous plot of $C(\alpha)$ and $S(\alpha)$ for different values of α. Now, we visualize a quantity $C(\alpha) + jS(\alpha)$ to be a complex phasor joining the origin to the point α on the spiral. Thus, $\{C(\zeta_2) - C(\zeta_1)\} + j\{S(\zeta_2) - S(\zeta_1)\}$ is the phasor defined by the line joining the point ζ_1 to the point ζ_2 on the spiral, so that $g(x)$, as defined by Eqs. (3.3-41), is the complex conjugate of this phasor multiplied by $(\pi z/k_0)^{1/2}$. We can compute $g(y)$ similarly and can numerically evaluate the Fresnel diffraction pattern.

Observe that at a very small distance from the aperture $(k_0 l^2/z \gg 1)$,

$$\zeta_{1,2} \to \begin{cases} -\infty & \text{for } x > \mp l/2, \\ +\infty & \text{for } x < \mp l/2, \end{cases}$$

$$\eta_{1,2} \to \begin{cases} -\infty & \text{for } y > \mp l/2, \\ +\infty & \text{for } y < \mp l/2, \end{cases} \qquad (3.3\text{-}43)$$

so that

$$C(\zeta_{1,2}) = S(\zeta_{1,2}) \to \begin{cases} -0.5 & \text{for } x > \mp l/2, \\ +0.5 & \text{for } x < \mp l/2, \end{cases}$$

$$C(\eta_{1,2}) = S(\eta_{1,2}) \to \begin{cases} -0.5 & \text{for } y > \mp l/2, \\ +0.5 & \text{for } y < \mp l/2. \end{cases} \qquad (3.3\text{-}44)$$

In Eqs. (3.3-43) and (3.3-44), the subscripts 1 and 2 in ζ and η, respectively corresponds to the $-$ and $+$ signs on the limits on x and y. Substituting Eqs. (3.3-44) into Eqs. (3.3-41) and, thereafter, the results into Eq. (3.3-38), we finally obtain

$$\psi_e(x, y, z) \simeq \text{rect}(x/l)\text{rect}(y/l), \qquad (3.3\text{-}45)$$

which implies that deep within the Fresnel region, the field distribution obeys the results predicted from geometrical optics, as expected.

Example 3.8 Plane Wave through a Circular Aperture

$$\psi_{e0} = \text{circ}(r/r_0) = \begin{cases} 1 & \text{for } r < r_0, \\ 0 & \text{otherwise.} \end{cases} \qquad (3.3\text{-}46)$$

It is difficult to find the general Fresnel diffraction pattern in this case; hence, we will limit ourselves to finding the on-axis intensity. (The Fraunhofer diffraction pattern is discussed later.) Again using Eq. (3.3-18),

$$\psi_e(0, 0, z) = \frac{jk_0}{2\pi z} \int_{-\infty}^{\infty} \int_{-\infty}^{\infty} \psi_{e0}(x', y') \exp\left[-j\left(\frac{k_0}{2z}\right)(x'^2 + y'^2)\right] dx' dy'$$

$$= \frac{jk_0}{2\pi z} \int_0^{r_0} \int_0^{2\pi} r \exp\left[-j\left(\frac{k_0}{2z}\right)r^2\right] dr\, d\theta$$

[changing to polar coordinates]

$$= 2j\left[\exp\left(-j\frac{k_0 r_0^2}{4z}\right)\right]\left[\sin\frac{k_0 r_0^2}{4z}\right]. \qquad (3.3\text{-}47)$$

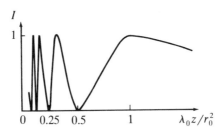

Figure 3.14 Variation of on-axis intensity for Fresnel diffraction from circular aperture.

Hence, the on-axis intensity is given by

$$I(0, 0, z) = 4\sin^2\left(\frac{N\pi}{2}\right),\qquad(3.3\text{-}48)$$

where N is the *Fresnel number* of the circular aperture, defined as

$$N = \frac{k_0 r_0^2}{2\pi z}.\qquad(3.3\text{-}49)$$

The intensity pattern is plotted in Figure 3.14. The intensity is zero where

$$z = z_m = \frac{k_0 r_0^2}{4m\pi}, \quad m \text{ integer},\qquad(3.3\text{-}50a)$$

and is a maximum where

$$z = z_M = \frac{k_0 r_0^2}{(4m + 2)\pi}, \quad m \text{ integer}.\qquad(3.3\text{-}50b)$$

Note that the distance between successive maxima keeps increasing as z increases, until after the point where $N = 1$ there is only a monotonic decrease in the on-axis intensity. Indeed, if we reexamine Eqs. (3.3-48) and (3.3-49), we conclude that for large z, the on-axis intensity decreases approximately according to $1/z^2$, which is consistent with the fact that from the *far field* the circular aperture approximates a point source.

To interpret the on-axis intensity variation derived above from a slightly different standpoint, note that the distance d between any point on the rim of the circular aperture and a point on-axis a distance z away is

$$d = \left[z^2 + r_0^2 \right]^{1/2} \simeq z + \frac{r_0^2}{2z}. \qquad (3.3\text{-}51)$$

Hence, the *path difference* between the center of the circular aperture and a point on the rim, as measured with respect to a point on-axis a distance z away, is

$$d - z = \frac{r_0^2}{2z}. \qquad (3.3\text{-}52)$$

If

$$d - z = \frac{n\lambda_0}{2} = \frac{n\pi}{k_0}, \quad n \text{ integer}, \qquad (3.3\text{-}53)$$

the aperture is said to contain n *half-period zones*. Note that Eq. (3.3-52) with Eq. (3.3-53) is identical to Eq. (3.3-50a) if $n = 2m$, and to Eq. (3.3-50b) if $n = 2m + 1$. The moral of the story is that the on-axis intensity will be a maximum or a minimum depending on whether there are an odd or even number of half-period zones. In fact, by blocking alternate half-period zones (as shown in Figure 3.15), one constructs what is called a *Fresnel zone plate*, which acts as a lens with multiple foci, the focal lengths being given by z_M.

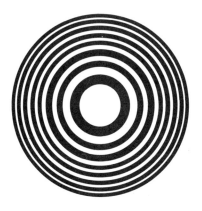

Figure 3.15 The Fresnel zone plate.

3.4 Huygens' Principle

From the Fresnel integral formula, Eq. (3.3-19), derived in the previous section, we can write

$$\psi_p = \frac{jk_0}{2\pi} \int_{-\infty}^{\infty}\int_{-\infty}^{\infty} \psi_{e0}(x', y')$$

$$\times \frac{1}{z}\exp\left(-jk_0\left[z + \frac{1}{2z}\{(x - x')^2 + (y - y')^2\}\right]\right) dx'\, dy'$$

$$= \frac{jk_0}{2\pi} \int_{-\infty}^{\infty}\int_{-\infty}^{\infty} \psi_{e0}(x', y')\frac{1}{R} \exp(-jk_0R)\, dx'\, dy', \qquad (3.4\text{-}1)$$

where $R = \{(x - x')^2 + (y - y')^2 + z^2\}^{1/2}$ represents the distance from every point $P'(x', y')$ on the input plane to a point $P(x, y)$ on the output (or observation) plane a distance z away from the input plane (see Figure 3.12), assuming $(x - x')_{\max}, (y - y')_{\max} \ll z$. Note that the term $(1/R)\exp(-jk_0R)$ in Eq. (3.4-1) is the phasor representation of a spherical wave propagating away from the input plane [see Eq. (3.3-26)]. Thus, Eq. (3.4-1) represents the contribution of all weighted spherical waves, or *Huygens wavelets*, from the input plane (x', y'). This is a consequence of *Huygens' principle*, which states that *every point on a primary wavefront serves as the source of spherical secondary wavelets such that the primary wavefront*

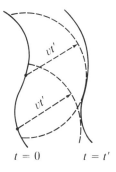

Figure 3.16 Propagation of a wavefront according to Huygens' principle.

at some later time is the envelope of these wavelets, as shown in Figure 3.16.

However, Huygens' principle, as stated above, is independent of any wavelength considerations when the wavefront passes through an aperture. This complication was taken care of by Fresnel in the *Huygens–Fresnel principle*, which states that *every unobstructed point of a wavefront, at a given instant in time, serves as a source of spherical secondary wavelets **of the same frequency** as the primary wave*. The amplitude of the optical field at any point beyond is the superposition of all of these wavelets, considering their relative amplitudes and phases. This, in fact, is what Eq. (3.4-1) states. The first mathematical formulation of the Huygens–Fresnel principle was done by Kirchhoff by solving the Helmholtz equation, Eq. (3.2-70), using Green's theorem [see, for instance, Goodman (1968)].

3.5 Fraunhofer Approximation and Fourier Optics

So far, we have studied the effect of propagation on the amplitude and phase distributions of various initial profiles. At all times, we examined the Fresnel diffraction pattern, which is determinable through the Fresnel diffraction formula, Eq. (3.3-18) or (3.3-19). The range of applicability of this formula is from distances not too close to the source (typically from about 10 times the wavelength). However, it is not always easy to determine the diffraction pattern, as was seen in the example of the rectangular aperture. In this section we examine a means of calculating the diffraction pattern at distances far away from the source or aperture. More precisely, observe that if our observation plane is in the *far field*, that is,

$$\frac{k_0(x'^2 + y'^2)_{\max}}{2z} = z_R \ll z, \qquad (3.5-1)$$

where z_R is the *Rayleigh range*, then the value of the exponential $[\exp - jk_0(x'^2 + y'^2)]_{\max}/2z$ is approximately unity over the input plane (x', y'). Under this assumption, which is commonly called the

Physical Optics

Fraunhofer approximation, Eq. (3.3-19) becomes

$$\psi_p = \frac{jk_0 e^{-jk_0 z}}{2\pi z} \exp\left[-j\frac{k_0}{2z}(x^2 + y^2)\right]$$

$$\times \int_{-\infty}^{\infty}\int_{-\infty}^{\infty} \psi_{e0}(x', y')\exp\left[j\frac{k_0}{z}(xx' + yy')\right] dx'\, dy'$$

$$= \frac{jk_0 e^{-jk_0 z}}{2\pi z} \exp\left[-j\frac{k_0}{2z}(x^2 + y^2)\right] \mathscr{F}_{xy}\{\psi_{e0}(x, y)\}\Big|_{\substack{k_x = k_0 x/z \\ k_y = k_0 y/z}}. \quad (3.5\text{-}2)$$

Equation (3.5-2) is termed the *Fraunhofer diffraction formula* and is the limiting case of the Fresnel diffraction studied earlier. The first exponential in Eq. (3.5-2) is the result of the phase change due to propagation, whereas the second exponential indicates a phase curvature that is quadratic in nature. Note that if we are treating diffraction of red light (λ_0 = 640 nm) and the maximum dimensions on the input plane are 1 mm, then $z \gg 5$ m. As a matter of fact, Fraunhofer diffraction can be observed at distances much smaller than the value just predicted. In what follows, we consider various examples of Fraunhofer diffraction when a plane wave passes through different apertures.

Example 3.9 Slit of Finite Width

$$\psi_{e0}(x, y) = \text{rect}\left(\frac{x}{l_x}\right). \quad (3.5\text{-}3)$$

Note that because we are usually interested in diffracted intensities (i.e., $|\psi_e|^2$ or $|\psi_p|^2$), the exponentials in Eq. (3.5-2) drop out. Furthermore, the other term besides the Fourier transform, namely, $(k_0/2\pi z)$, simply acts as a weighting factor. The intensity profile depends on the Fourier transform, and we will therefore concentrate only on this unless otherwise stated.

At this point, we also caution readers on the use of the Fraunhofer diffraction formula. In our problem, note that there is

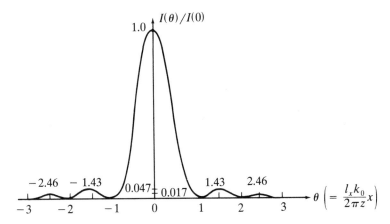

Figure 3.17 Fraunhofer diffraction pattern of a rectangular slit.

no obstruction to the incident plane wave in the y dimension when it emerges from the aperture. Thus, there will not be any effect of diffraction in the y direction. We only need to take the Fourier transform in x to find the Fraunhofer diffraction pattern. Using Eq. (3.5-3),

$$\mathscr{F}_x\left\{\text{rect}\left(\frac{x}{l_x}\right)\right\} = l_x \, \text{sinc}\left(\frac{l_x k_x}{2\pi}\right) \qquad (3.5\text{-}4)$$

(see Table 1.3). Hence, using Eq. (3.5-2),

$$\psi_p \propto l_x \, \text{sinc}\left(\frac{l_x k_0 x}{2\pi z}\right). \qquad (3.5\text{-}5)$$

Observe that the first zero of the sinc function occurs at $x = \pm 2\pi z/l_x k_0 = \pm \lambda_0 z/l_x$, and it is between these points that most of the diffracted intensity ($\propto |\psi_p|^2$) falls (see Figure 3.17). This is to be expected, because the heuristic treatment of diffraction in Section 3.3 predicts a diffraction angle of the order of λ_0/l_x.

Example 3.10 Rectangular Aperture

$$\psi_{e0}(x, y) = \text{rect}\left(\frac{x}{l_x}\right)\text{rect}\left(\frac{y}{l_y}\right). \tag{3.5-6}$$

Following the results of the previous example,

$$\psi_p \propto l_x l_y \, \text{sinc}\left(\frac{l_x k_0 x}{2\pi z}\right)\text{sinc}\left(\frac{l_y k_0 y}{2\pi z}\right) \tag{3.5-7}$$

and is the far-field approximation to the Fresnel diffraction pattern of the rectangular aperture that we studied in Section 3.3.

Example 3.11 Young's Double-Slit Experiment

$$\psi_{e0} = \left[\delta\left(x - \frac{d}{2}\right) + \delta\left(x + \frac{d}{2}\right)\right]\delta(y). \tag{3.5-8}$$

(See Figure 3.18.) Young's double-slit experiment reveals the fundamentals of *interference*, which can be viewed also as an example

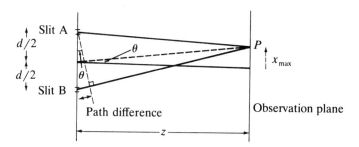

Figure 3.18 Young's double-slit experiment. The diagram also shows how to calculate the path difference between the slits to the observation point.

of diffraction. In this case,

$$\psi_p \propto \mathscr{F}_{xy}\left[\left\{\delta\left(x - \frac{d}{2}\right) + \delta\left(x + \frac{d}{2}\right)\right\}\delta(y)\right]\Bigg|_{k_x=k_0x/z,\,k_y=k_0y/z}$$

$$= \left(\exp\left[-\frac{jk_xd}{2}\right] + \exp\left[\frac{jk_xd}{2}\right]\right)\Bigg|_{k_x=k_0x/z}$$

$$\propto \cos\left(\frac{k_0x}{2z}d\right). \tag{3.5-9}$$

The intensity along x is proportional to $\cos^2((k_0d/2z)x)$; hence, we expect maxima in the intensity profile at

$$x_{max} = \frac{2n\pi}{k_0d}z = n\left(\frac{\lambda_0}{d}\right)z, \qquad n \text{ integer.} \tag{3.5-10}$$

Note the presence of the parameter λ_0/d, which is proportional to the angle of diffraction.

Alternatively, we can derive the locations of the maxima by comparing the path lengths that waves emanating from the two slits travel to reach a point P on the observation plane (see Figure 3.18), and setting the difference between these to be an integral number of wavelengths for what is called *constructive interference*. From the geometry in Figure 3.18 it is clear that

$$\theta \sim \frac{\text{path difference}}{d} = \frac{x_{max}}{z}, \qquad \theta \text{ small.} \tag{3.5-11}$$

Now, setting the path difference equal to $n\lambda_0$, Eq. (3.5-10) follows.

Example 3.12 Circular Aperture

$$\psi_{e0}(x, y) \triangleq \psi_{e0}(r) = \text{circ}\left(\frac{r}{r_0}\right). \tag{3.5-12}$$

Note that $r = (x^2 + y^2)^{1/2}$ and that $\text{circ}(r/r_0)$ denotes a value 1 within a circle of radius r_0 and 0 otherwise. The first step is to

Physical Optics

compute the Fourier transform of the circ function. Now,

$$\mathscr{F}_{xy}\{\psi_{e0}\} = \Psi_{e0}(k_x, k_y) \triangleq \int_{-\infty}^{\infty}\int_{-\infty}^{\infty} \psi_{e0}(x, y)\exp\left[j(k_x x + k_y y)\right] dx\, dy.$$

(3.5-13)

Introducing new variables

$$x = r\cos\theta, \qquad y = r\sin\theta, \qquad k_x = k_r\cos\phi, \qquad k_y = k_r\sin\phi,$$

(3.5-14)

Eq. (3.5-13) transforms to

$$\Psi_{e0}(k_x, k_y) \triangleq \overline{\Psi}_{e0}(k_r, \phi)$$

$$= \int_{-\infty}^{\infty}\int_0^{2\pi} \overline{\psi}_{e0}(r)\exp[jk_r r(\cos\theta\cos\phi$$

$$+ \sin\theta\sin\phi)]r\, d\theta\, dr$$

$$= \int_0^{\infty} r\overline{\psi}_{e0}(r)\int_0^{2\pi} \exp[jk_r r\cos(\theta - \phi)]\, d\theta\, dr, \quad (3.5\text{-}15)$$

where we have employed the circular symmetry of $\psi_{e0}(x, y) = \overline{\psi}_{e0}(r)$. But,

$$J_0(\beta) \triangleq \frac{1}{2\pi}\int_0^{2\pi} \exp[j\beta\cos(\theta - \phi)]\, d\theta, \qquad (3.5\text{-}16)$$

where $J_0(\beta)$ is the zeroth-order Bessel function. [Note that the LHS of Eq. (3.5-15) is independent of ϕ.] Using Eq. (3.5-16) in Eq. (3.5-15), we obtain

$$\mathscr{B}\{\overline{\psi}_{e0}\} \triangleq \overline{\Psi}_{e0}(k_r) = 2\pi\int_0^{\infty} r\overline{\psi}_{e0}(r)J_0(k_r r)\, dr. \quad (3.5\text{-}17)$$

Equation (3.5-17) defines the *Fourier–Bessel transform* and arises in circularly symmetric problems. Now, substituting Eq. (3.5-12) with

(3.3-46) in Eq. (3.5-17), we get

$$\overline{\Psi}_{e0}(k_r) = 2\pi \int_0^{r_0} rJ_0(k_r r)\, dr. \qquad (3.5\text{-}18)$$

Finally, using the result

$$\alpha J_1(\alpha) = \int_0^{\alpha} \beta J_0(\beta)\, d\beta, \qquad (3.5\text{-}19)$$

Eq. (3.5-18) becomes

$$\overline{\Psi}_{e0}(k_r) = \frac{2\pi r_0}{k_r} J_1(r_0 k_r), \qquad (3.5\text{-}20)$$

which is the desired Fourier transform. Hence, from Eq. (3.5-2), we
have

$$\psi_p \propto \mathscr{F}_{xy}\{\psi_{e0}\}\Big|_{\substack{k_x=k_0 x/z \\ k_y=k_0 y/z}}$$

$$= \mathscr{B}\{\overline{\psi}_{e0}\}\Big|_{k_r=k_0 r/z}$$

or

$$\overline{\psi}_p(r,z) \triangleq \psi_p(x,y,z) \propto \frac{2\pi r_0 z}{k_0 r} J_1\!\left(\frac{r_0 k_0}{z} r\right) \qquad (3.5\text{-}21)$$

The intensity is proportional to $|\overline{\psi}_p|^2$ and is sketched in Figure 3.19.
The plot is called an *Airy pattern*.

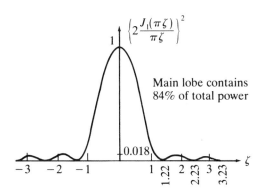

Figure 3.19 Plot of $[2J_1(\pi\zeta)/\pi\zeta]^2$ against ζ $(= r_0 k_0 r/\pi z)$. This represents
the Fraunhofer diffraction pattern of the circular aperture.

The circular aperture is of special importance in optics because lenses, which form an important part in any optical system, usually have a circular cross section. In what follows, we derive the *Rayleigh's criterion*, which dictates the *resolution* of an optical imaging system. Resolution is a figure of merit that determines how close two object points can be such that they are clearly distinguished, or *resolved*, by the optical system. Note from Figure 3.19 that most of the intensity lies between $|\zeta| \leq 1.22$. Consider, now, two Airy patterns superposed on each other with a certain distance of separation between their peaks. It is easy to understand that the peaks of the main lobes should be no closer than 1.22 units (in ζ) for them to be discernible in the superposed picture. In the context of our problem, this translates to

$$\frac{r_0 k_0}{z} r \geq 1.22\pi$$

or

$$\alpha_{min} \triangleq \left(\frac{r}{z}\right)_{min} = \frac{(1.22)\pi}{k_0 r_0} = 0.61\left(\frac{\lambda_0}{r_0}\right), \qquad \alpha_{min} \text{ small}, \quad (3.5\text{-}22)$$

where α_{min} represents the minimum angle between the beams contributing to the two main lobes to facilitate resolution. Note, yet again, the omnipresence of the parameter λ_0/r_0.

To see where the Rayleigh criterion, as enunciated in Eq. (3.5-22), plays an important role in imaging systems, consider the arrangement shown in Figure 3.20, where we have two point sources, P_1 and P_2, a distance x_0 apart on the front focal plane of a lens of

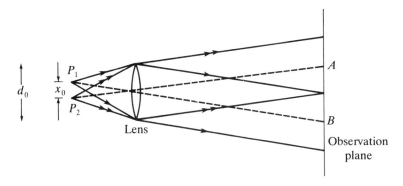

Figure 3.20 Optical arrangement to derive the Abbe condition.

focal length f. From elementary ray optics, which we studied in Chapter 2 (and also from the wave optics of lenses, as we shall see in the next chapter), two beams should propagate on the right-hand side of the lens, at an angle to each other. The beam diameter is, however, finite, due to the finite diameter d_0 of the lens. In fact, the lens of finite diameter can be replaced by one with, hypothetically, an infinite diameter, followed by a circular aperture of diameter d_0. In our case, the circular aperture is illuminated by plane wavefronts, because a point source on the front focal plane of a lens of infinite aperture produces plane waves behind it. Thus, on the observation plane, we can see the diffraction patterns of the aperture. Looking from the observation plane, we can therefore distinguish P_1 and P_2 as long as the angle between the two beams is more than α_{min}. From the geometry of Figure 3.20 it follows that the minimum separation $(x_0)_{min}$ between P_1 and P_2 is determined from

$$\frac{(x_0)_{min}}{f} \simeq \alpha_{min} = 1.22\left(\frac{\lambda_0}{d_0}\right),$$

or

$$(x_0)_{min} \simeq 1.22\lambda_0\left(\frac{f}{d_0}\right). \qquad (3.5\text{-}23)$$

This criterion is often called the *Abbe condition*.

The problem discussed above can be switched around, by asking for the minimum angle between two beams starting from sources A and B at infinity and passing through the lens such that their images can be resolved or, equivalently, by inquiring about the minimum separation between their peaks. The answer, found conventionally in optics texts, is identical to Eq. (3.5-23).

The parameter f/d_0 is the *f-number* of the lens. For instance, a lens with a 25-mm aperture and a focal length of 50 mm has an *f*-number equal to 2 and is designated as $f/2$. Cameras have a fixed lens but a variable aperture (*diaphragm*) with typical *f*-number markings of 2, 2.8, 4, 5.6, 8, 11, and so forth. Each consecutive diaphragm setting increases the *f*-number by $\sqrt{2}$; hence, the amount of light energy reaching the film for the same exposure time is cut in half because this is proportional to the area of the beam and, hence, to $1/(f\text{-number})^2$. Note also that for the same amount of light energy to reach the film, the product of the exposure time and irradiance must be a constant; hence, the exposure time is proportional to $(f\text{-number})^2$ in this case. Logically, the f-number is sometimes called the *speed* of the lens. Thus an $f/1.4$ lens is said to be twice as *fast* as an $f/2$ lens. We will discuss films in greater detail in Chapter 5.

Example 3.13 Sinusoidal Amplitude Grating

$$\psi_{e0}(x, y) = \left\{ \frac{1}{2} + \frac{m}{2} \cos(k_{x_0} x) \right\} \text{rect}\left(\frac{x}{l} \right) \text{rect}\left(\frac{y}{l} \right), \qquad m < 1.$$

$$(3.5\text{-}24)$$

To find $\psi_p(x, y, z)$, note that

$$\mathcal{F}_{xy}\left\{ \frac{1}{2} + \frac{m}{2} \cos(k_{x_0} x) \right\} = \frac{1}{2}\delta(k_x, k_y) + \frac{m}{2}\delta(k_x + k_{x_0}, k_y)$$

$$+ \frac{m}{2}\delta(k_x - k_{x_0}, k_y), \qquad (3.5\text{-}25a)$$

and

$$\mathcal{F}_{xy}\left\{ \text{rect}\left(\frac{x}{l} \right) \text{rect}\left(\frac{y}{l} \right) \right\} = l^2 \, \text{sinc}\left(\frac{l k_x}{2\pi} \right) \text{sinc}\left(\frac{l k_y}{2\pi} \right). \quad (3.5\text{-}25b)$$

We use the frequency convolution theorem to calculate the Fourier transform of Eq. (3.5-24), using Eq. (3.5-25). Analogous to the procedure followed in finding the Fourier transform of the convolution of two spatial functions (see Chapter 1), we can show that

$$\mathcal{F}_{xy}[g_1(x, y)g_2(x, y)] = \frac{1}{4\pi^2}[G_1(k_x, k_y) * G_2(k_x, k_y)], \quad (3.5\text{-}26)$$

where G_1 and G_2 are the Fourier transforms of g_1 and g_2,

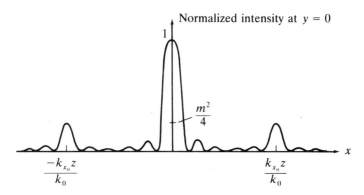

Figure 3.21 Intensity distribution for Fraunhofer diffraction from an amplitude grating.

respectively. Next, upon replacing k_x and k_y by $k_0 x/z$ and $k_0 y/z$, respectively, we get the electric field, and the square of its absolute value yields the intensity. This is plotted in Figure 3.21.

3.6 Wave Propagation in an Inhomogeneous Medium: Graded-Index Fiber

Thus far, we have only considered wave propagation in a homogeneous medium, characterized by a constant permittivity ϵ. In inhomogeneous materials, the permittivity can be a function of the spatial coordinates x, y, and z. To study wave propagation in inhomogeneous materials, we need to return to Maxwell's equations [Eqs. (3.1-1)–(3.1-4)] and rederive the wave equation. Our starting point is Eq. (3.2-4), which we rewrite here for a source-free medium ($\mathbf{J}_c = 0$):

$$\nabla^2 \mathbf{E} - \mu\epsilon \frac{\partial^2 \mathbf{E}}{\partial t^2} = \nabla(\nabla \cdot \mathbf{E}). \qquad (3.6\text{-}1)$$

Now, from Eq. (3.1-1), with $\rho = 0$, and Eq. (3.1-13a),

$$\nabla \cdot (\epsilon \mathbf{E}) = \epsilon \nabla \cdot \mathbf{E} + \mathbf{E} \cdot \nabla\epsilon = 0. \qquad (3.6\text{-}2)$$

Physical Optics

With Eq. (3.6-2), we can rewrite Eq. (3.6-1) as

$$\nabla^2 \mathbf{E} - \mu\epsilon \frac{\partial^2 \mathbf{E}}{\partial t^2} + \nabla\left(\mathbf{E} \cdot \frac{\nabla\epsilon}{\epsilon}\right) = 0. \qquad (3.6\text{-}3)$$

If the spatial variation of ϵ is small over a wavelength of the propagating field, we can neglect the last term on the LHS of Eq. (3.6-3) to write

$$\nabla^2 \mathbf{E} - \mu\epsilon \frac{\partial^2 \mathbf{E}}{\partial t^2} = 0, \qquad (3.6\text{-}4)$$

where $\epsilon = \epsilon(x, y, z)$. Note that Eq. (3.6-4) is similar to the homogeneous wave equation for the electric field, Eq. (3.2-10), derived earlier. For notational convenience, we return to our generic dependent variable $\psi(x, y, z, t)$ and adopt

$$\nabla^2 \psi - \mu_0\epsilon \frac{\partial^2 \psi}{\partial t^2} = 0, \qquad \epsilon = \epsilon(x, y, z), \qquad (3.6\text{-}5)$$

as our model equation, where we have assumed $\mu = \mu_0$ for simplicity.

We will restrict ourselves to inhomogeneous media like graded-index optical fibers. Recall from Section 2.3 that the refractive index profile in such a profile can be modelled as in Eq. (2.3-6). Equivalently, we can incorporate the inhomogeneity through a permittivity profile of the form

$$\epsilon(x, y, z) = \epsilon(x, y) = \epsilon(0)\left\{1 - \frac{x^2 + y^2}{h^2}\right\}, \qquad (3.6\text{-}6)$$

where we have assumed two transverse dimensions instead of one as in Eq. (2.3-6).

We wish to study the propagation of arbitrary beam profiles through the inhomogeneous medium modelled by Eq. (3.6-6). However, solution of Eq. (3.6-5) with Eq. (3.6-6) is difficult for arbitrary initial conditions. We therefore look for a propagating solution of a single frequency ω_0 that can have an arbitrary cross-sectional amplitude and/or phase profile. Thus, we set

$$\psi(x, y, z, t) = \text{Re}[\psi_e(x, y)\exp(j(\omega_0 t - kz))] \quad (3.6\text{-}7)$$

and substitute into the wave equation, Eq. (3.6-5). This gives

$$\nabla_T^2 \psi_e + [\omega_0^2 \mu_0 \epsilon(x, y) - k^2]\psi_e = 0, \quad (3.6\text{-}8)$$

where ∇_T^2 denotes the *transverse Laplacian* $\partial^2/\partial x^2 + \partial^2/\partial y^2$. We denote by k_0 the propagation constant of an infinite plane wave propagating in a medium of uniform dielectric constant $\epsilon(0)$, that is,

$$k_0 = \omega_0(\mu_0 \epsilon(0))^{1/2}. \quad (3.6\text{-}9)$$

When we introduce Eq. (3.6-6) into Eq. (3.6-8) with Eq. (3.6-9) and use the normalized variables

$$\xi = \left(\frac{k_0}{h}\right)^{1/2} x, \qquad \eta = \left(\frac{k_0}{h}\right)^{1/2} y, \quad (3.6\text{-}10)$$

we get

$$\frac{\partial^2 \overline{\psi}_e}{\partial \xi^2} + \frac{\partial^2 \overline{\psi}_e}{\partial \eta^2} + [\lambda - (\xi^2 + \eta^2)]\overline{\psi}_e = 0, \qquad \overline{\psi}_e(\xi, \eta) = \psi_e(x, y),$$

$$(3.6\text{-}11)$$

where

$$\lambda = \frac{(k_0^2 - k^2)h}{k_0}. \quad (3.6\text{-}12)$$

We solve Eq. (3.6-11) using the commonly used *separation of variables* technique. To this end, we assume $\overline{\psi}_e(\xi, \eta) = X(\xi)Y(\eta)$, substitute in Eq. (3.6-11), and derive two decoupled ordinary differential equations (ODEs) for X and Y:

$$\frac{d^2 X}{d\xi^2} + (\lambda_x - \xi^2)X = 0, \quad (3.6\text{-}13a)$$

$$\frac{d^2 Y}{d\eta^2} + (\lambda_y - \eta^2)Y = 0, \quad (3.6\text{-}13b)$$

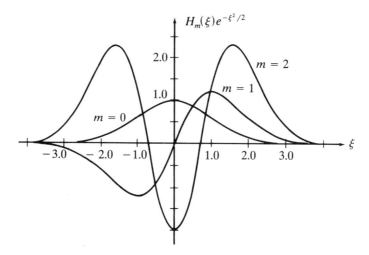

Figure 3.22 The three lowest-order Hermite–Gaussian functions.

with $\lambda_x + \lambda_y = \lambda$. Each of Eqs. (3.6-13a) and (3.6-13b) is of the same form as that arising in the analysis of the harmonic oscillator problem in quantum mechanics [Schiff (1968)]. The solution to Eq. (3.6-13a) is

$$X_m(\xi) = H_m(\xi)e^{-\xi^2/2}, \quad \lambda_x = 2m + 1, \, m = 0, 1, 2, \ldots, \quad (3.6\text{-}14)$$

where the H_ms are called the *Hermite polynomials*. The first few Hermite polynomials are

$$H_0(\xi) = 1, \quad H_1(\xi) = 2\xi, \quad H_2(\xi) = 4\xi^2 - 2, \ldots. \quad (3.6\text{-}15)$$

The solutions to $X(\xi)$ are called *Hermite–Gaussians*. The first few are plotted in Figure 3.22. Similar solutions hold for $Y(\eta)$. Equation (3.6-11) thus has the general solution

$$\bar{\psi}_e(\xi, \eta) \triangleq \bar{\psi}_{e_{mn}}(\xi, \eta) = H_m(\xi)H_n(\eta)\exp\left[-\frac{(\xi^2 + \eta^2)}{2}\right], \quad (3.6\text{-}16)$$

with

$$\lambda \triangleq \lambda_{mn} = 2(m + n) + 2, \quad m, n = 0, 1, 2, \ldots. \quad (3.6\text{-}17)$$

$\bar{\psi}_{e_{mn}}$ is called the *mode pattern* or *mode profile* of the mnth mode.

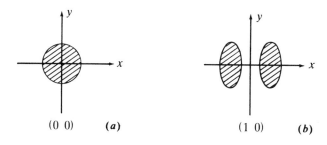

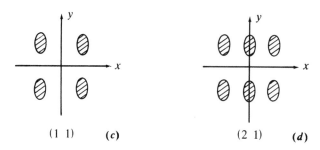

Figure 3.23 Intensity patterns of light in the 00, 10, 11, and 21 modes during propagation through a graded-index fiber with a quadratic refractive index (and permittivity) profile.

The fundamental mode, with $m = n = 0$ ($\lambda_{00} = 2$) is given by

$$\psi_{e_{00}}(x, y) = \frac{(2/\pi)^{1/2}}{w} \exp\left[\frac{-(x^2 + y^2)}{w^2}\right], \qquad (3.6\text{-}18)$$

where

$$w = \left(\frac{2h}{k_0}\right)^{1/2}. \qquad (3.6\text{-}19)$$

The corresponding intensity pattern is shown in Figure 3.23(a).

Intensity patterns corresponding to higher-order modes are shown in Figures 3.23(b)–3.23(d). The propagation constant $k \triangleq k_{mn}$ of the mnth mode is obtained from Eq. (3.6-12) with Eq. (3.6-17):

$$k^2_{mn} = k^2_0 \left[1 - \frac{2(m + n + 1)}{k_0 h} \right], \qquad (3.6\text{-}20)$$

indicating that the propagation constant decreases with increasing mode number. Thus, higher-order modes have higher phase velocities and are composed of waves with $\mathbf{k}$ vectors of greater angles since these mode profiles extend further away from the z-axis. Due to the finite radius of the fiber, which is an example of an inhomogeneous medium, it turns out that a few of the lowest-order Hermite–Gaussians can really exist. More on the Hermite–Gaussian modes during optical propagation appears in Chapter 6 in connection with the modes in an optical resonator.

The preceding analysis is indicative of the mode patterns that are characteristic of a multimode fiber. An arbitrary excitation at the input of the fiber can be tracked by decomposing it into the characteristic modes just discussed. The Hermite–Gaussian functions form an orthogonal basis, enabling such a decomposition to be made easily. We also comment that multimode fibers possess a distinct disadvantage in the sense that different modes travel with different velocities, leading to *modal dispersion*. Optical pulses travelling through multimode fibers are more easily dispersed or spread in time than in single-mode fibers, which are more commonly used for optical communication. Single-mode fibers are usually constructed with a step-index geometry, where the core of the fiber has a constant refractive index that is higher than that of the cladding. The refractive indices are chosen in such a way that outside the core, only evanescent solutions exist and that only one mode, namely, the zeroth order, can propagate inside the core.

Problems

3.1　For a time-harmonic uniform plane wave in a linear isotropic homogeneous medium, the $\mathbf{E}$ and $\mathbf{H}$ fields vary according to $\exp(\omega_0 t - \mathbf{k}_0 \cdot \mathbf{R})$. Show that in this case, Maxwell's equations in a source-free region can be expressed as

$$\mathbf{k}_0 \cdot \mathbf{E} = 0,$$

$$\mathbf{k}_0 \cdot \mathbf{H} = 0,$$

$$\mathbf{k}_0 \times \mathbf{E} = \omega_0 \mu \mathbf{H},$$

$$\mathbf{k}_0 \times \mathbf{H} = -\omega_0 \epsilon \mathbf{E}.$$

3.2 Verify Eq. (3.2-23) by direct substitution into the wave equation,

$$\frac{\partial^2 \psi}{\partial R^2} + \frac{2}{R} \frac{\partial \psi}{\partial R} = \left(\frac{k_0}{\omega_0} \right)^2 \frac{\partial^2 \psi}{\partial t^2}.$$

3.3 Consider three plane waves travelling in three different directions, as shown in Figure P3-3.
(a) Assuming that they all start in phase at the point O and have the same frequency, calculate the intensity $I(y)$ at the point P on the observation plane in terms of the relative field strength a, the angle ϕ, and the distances y_0 and z_0.
(b) From this determine the values of z_0 where one can observe maximum and minimum fringe contrasts. (Fringe contrast is $[I(y)]_{\text{max}}/[I(y)]_{\text{min}}$.)

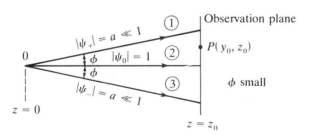

Figure P3.3

3.4 **(a)** Write down the Jones vector for circularly polarized light. Also, find a polarization state that is orthogonal to this state. Describe physically the differences between these two states. [*Hint:* Two complex vectors **A** and **B** are orthogonal if $(\mathbf{A}^*)^T \mathbf{B} = 0$.]

(b) Determine the Jones matrix that will facilitate transformation between the states in part (a).

3.5 Starting from Eq. (3.2-42) and employing the electromagnetic boundary conditions, prove Eqs. (3.2-47) and (3.2-48).

3.6 Show that when the angle of incidence exceeds the critical angle at the interface between two semiinfinite dielectrics,

$$|r_\perp| = |r_\parallel| = 1.$$

3.7 In Figure 3.8, the sharpness of the transmission peaks is determined by the half-width $\delta_{1/2}$, which is defined as the value of δ off δ_{max} (where δ_{max} is the value of δ corresponding to a transmission peak) where the intensity becomes equal to half the peak value. Show that

$$\delta_{1/2} = 2\sin^{-1}(1/\sqrt{F}).$$

3.8 A dispersive medium is characterized by the dispersion relation

$$\omega^2 = c^2 k^2 + \omega_c^2, \qquad \omega > 0.$$

(a) Write down the dispersion relation for $\omega < 0$.
(b) Calculate the phase and group velocities,
(c) Determine the PDE for the real wave function ψ travelling through the medium.

3.9 Verify that in the eikonal equations, Eqs. (3.2-74) and (3.2-75), the last term on the RHS of Eq. (3.2-74) represents the contribution due to diffraction. (*Hint:* To check this, delete this term and solve the set of equations to show that an arbitrary amplitude profile is unchanged in shape during propagation. Assume $\phi = k_{x0}x + k_{y0}y + k_{z0}z$.)

3.10 Find a paraxial approximation to a wavefront, in the plane $z = 0$, that converges to the point P as shown in Figure P3.10. A transparency $t(x, y)$ is now placed at $z = 0$ and illuminated by the converging wavefront. Assuming Fresnel diffraction from $z = 0$ to the plane $z = z_0$, find the intensity pattern on the observation plane. Comment on the usefulness of illuminating an aperture or transparency using a spherical wavefront instead of a plane wavefront.

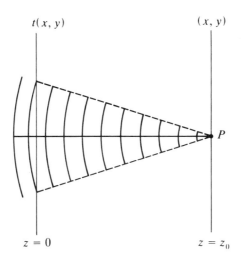

Figure P3.10

3.11 Find the *Fresnel* diffraction pattern of an initial amplitude profile

$$\psi_{e0}(x, y) = \left[\delta\left(x - \frac{x_0}{2}\right) + \delta\left(x + \frac{x_0}{2}\right)\right]\delta(y).$$

3.12 Consider the propagation of an elliptical Gaussian beam profile described by

$$\psi_{e0}(x, y) = \left[\exp\left\{-\left(\frac{x}{w_{0x}}\right)^2\right\}\right]\left[\exp\left\{-\left(\frac{y}{w_{0y}}\right)^2\right\}\right].$$

Show that Eqs. (3.3-34) and (3.3-35) describing the waist and the radius of curvature may be applied separately to the x and y dependences of the elliptic Gaussian.

3.13 Verify that Eq. (3.3-45) is the Fresnel diffraction pattern of a square aperture of side l, at a distance $z \ll k_0 l^2$ away, by first confirming Eqs. (3.3-43) and (3.3-44).

3.14 Determine the Fresnel diffraction pattern of a straight edge $u(x)$, where $u(x)$ denotes the unit step function, illuminated by a plane wavefront. Plot the intensity distribution. Reconcile your answer with what you would expect in the very near and far fields, respectively.

3.15 Find the on-axis intensity variation when a circular aperture of radius r_0 is illuminated by a point source located at a distance z_0 in front of the aperture. Compare your result with that in Example 3.8 in the text.

3.16 In Eq. (3.5-7), take the limit as $l_x, l_y \to \infty$. Do your results make physical sense? Give reasons for any discrepancy.

3.17 Find the Fraunhofer diffraction pattern of two slits described by the transparency function,

$$t(x, y) = \text{rect}\left[\frac{x - X/2}{x_0}\right] + \text{rect}\left[\frac{x + X/2}{x_0}\right],$$

in two ways:
(a) by directly taking the Fourier transform;
(b) by using the notion of convolution and then using the Fourier transform.
Suppose that the slits are now illuminated by light containing two frequencies ω_1 and ω_2. Suppose also that in the Fraunhofer diffraction pattern, the second off-axis maximum of color ω_1 coincides with the third off-axis maximum of color ω_2. Find the ratio of the two frequencies (ω_1/ω_2) constituting the illuminating light.

3.18 A plane wave of amplitude A propagating in the $+z$ direction is incident on an infinite series of slits, spaced S apart and a wide at $z = 0$, as shown in Figure P3.18. (Such a grating is called a *Ronchi grating*.) Find an expression for the amplitude distribution in the Fraunhofer diffraction of the aperture. Sketch the intensity distribution on the observation plane, labelling the coordinates of any important points.

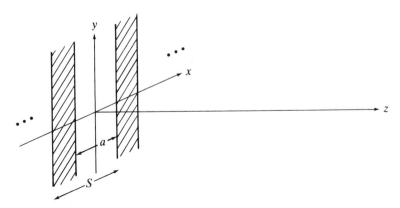

Figure P3.18

3.19 Find the Fraunhofer diffraction pattern of a sinusoidal phase grating described by $\exp[j(m/2)\sin(k_{x0}x)]\text{rect}(x/l)\,\text{rect}(y/l)$. Describe qualitatively what you would expect if the phase grating is moving along x with a phase velocity $V = \Omega/k_{x0}$.

3.20 Show the orthogonality property of the Hermite–Gaussian modes propagating through a graded-index fiber by proving that

$$\int_{-\infty}^{\infty} X_m(\xi)X_n(\xi)\,d\xi = 0 \quad \text{for } m \neq n,$$

where $X_m(\xi) = H_m(\xi)e^{-\xi^2/2}$. [*Hint:* Use the ODE for the Hermite–Gaussians as in Eq. (3.6-13) as a starting point.]

References

3.1 Banerjee, P. P. (1985). *Proc. IEEE* **73** 1859–1860.
3.2 Born, M. and E. Wolf (1983). *Principles of Optics*. Pergamon, New York.
3.3 Cheng, D. K. (1983). *Field and Wave Electromagnetics*. Addison-Wesley, Reading, Massachusetts.
3.4 Ghatak, A. K. and K. Thygarajan (1978). *Contemporary Optics*. Plenum, New York.

3.5 Goodman, J. W. (1968). *Introduction to Fourier Optics.* McGraw-Hill, New York.

3.6 Hecht, E. and A. Zajac (1975). *Optics.* Addison-Wesley, Reading, Massachusetts.

3.7 Korpel, A. and P. P. Banerjee (1984). *Proc. IEEE* **72** 1109.

3.8 Lee, D. L. (1986). *Electromagnetic Principles of Integrated Optics.* Wiley, New York.

3.9 Schiff, L. I. (1968) *Quantum Mechanics.* McGraw-Hill, New York.

3.10 Yu, F. T. S. (1983). *Optical Information Processing.* Wiley, New York.

Chapter 4   Optical Information Processing (I)

It is common practice in communications and signal processing to use *temporal* Fourier transforms in the study of input–output relationships of linear time-invariant systems. The output in the time domain is related to the input through a convolution integral. As we have seen in the previous chapter, *spatial* Fourier transforms were used to describe propagational diffraction effects in a medium. The angular plane-wave spectrum at the output plane is equal to the product of the spectrum at the input plane and the spatial transfer function of propagation. In the spatial domain, this means that the input and the output are once again related through a convolution integral, which is the Fresnel diffraction formula.

To carry this comparison one step further, note that in signal processing, we can design (temporal) filters to extract the relevant information from an input signal once we know its (temporal) spectrum. Similarly, to process optical information, such as images, we need to devise an efficient optical Fourier-transformer and suitable spatial filters. Of course, we know from Chapter 3 that the far-field diffraction pattern of a transparency illuminated by a plane wave is proportional to the spatial Fourier transform of the transparency. However, there are two major disadvantages in trying to utilize the spatial Fourier transform for image processing. First, we have to work in the far-field regime, which is not space-efficient.

Second, and more serious, we do not have an exact Fourier transform in the far field in the sense that the Fourier transform is premultiplied by a quadratic phase curvature [see Eq. (3.5-2)]. This poses complications if we want to perform spatial filtering on the Fourier transform.

As we shall see in this chapter, lenses provide a convenient means to bring the spatial Fourier transform of a transparency from the far field to the focal plane of the lens. Furthermore, if the transparency is placed in the front focal plane of a converging lens, we get an exact spatial Fourier transform on the back focal plane *without any phase distortions*. We also examine image formation in the diffraction-limited case. We recover the celebrated thin-lens formula that was derived in Chapter 2 using the geometrical optics treatment. We also study simple image processing systems and develop the concept of the (spatial) frequency response of imaging systems under coherent illumination, for example, laser light. As we shall show, the nature of a processed image is easily predictable from the geometrical image and the characteristics of the spatial filter. Furthermore, we discuss image processing using incoherent light, for example, the light from a sodium vapor lamp, mainly to point out the essential differences from coherent imaging systems. (Chapter 5 is devoted to more complex image processing systems, construction and use of complex spatial filters, and holography.)

4.1 Fourier-Transforming Property of Lenses

The advantage of having a spherical wavefront instead of a plane wavefront illuminating an aperture or transparency is appreciated from the results of Problem 3–10. The results indicate that by using converging spherical wave illumination of a transparency, we can bring the Fourier transform plane from the far field to the plane containing the point where the spherical wave converges. To utilize this advantage, however, the transformation of a plane wavefront to a spherical wavefront is essential. We can achieve this transformation by using optical elements such as lenses and mirrors. In this section, we study how the lens, which is a phase object, can perform this transformation. We will treat the transformation by mirrors briefly in Section 4.4.

In what follows, we first find the optical field immediately behind a phase object, such as a block of glass of varying thickness, when illuminated by a uniform plane wave. We then find the field

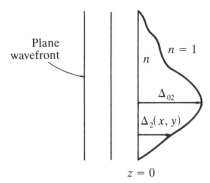

Figure 4.1 Plane wave fronts incident on a block of optical dense material of varying thickness.

distribution at an arbitrary distance. Thereafter, we determine the required nature of the phase object to effect the phase transformation from a plane wavefront to a spherical wavefront and find the point of convergence of these spherical waves. The spherical phase object turns out to be a lens, and the convergence point is called the *focal point*. Next, for an arbitrary transparency–lens combination, we determine the field distribution in the focal plane of the lens. We find that this field always contains the information of the Fourier transform of the transparency function.

Consider, therefore, a block of glass in air, of varying thickness $\Delta_2(x, y)$ and having a maximum thickness of Δ_{02}, as shown in Figure 4.1. If a plane wave of unit amplitude is incident from the left-hand side as shown, it suffers a varying phase delay,

$$\phi(x, y) = nk_0\Delta_2(x, y) + k_0[\Delta_{02} - \Delta_2(x, y)], \quad (4.1\text{-}1)$$

during travel from the input plane $z = 0$ to the plane $z = \Delta_{02}$. In Eq. (4.1-1), n is the refractive index of the glass. Also, $k_0 = \omega_0/c$ is the propagation constant for light of angular frequency ω_0 in air or vacuum. The first term on the RHS in Eq. (4.1-1) denotes the phase change in the glass, whereas the second term is the contribution due to the phase change in air between $z = \Delta_2$ and $z = \Delta_{02}$. As a check, if $\Delta_2(x, y) = \Delta_{02}$ (uniform slab), $\phi(x, y) = nk_0\Delta_{02}$, as expected.

Optical Information Processing (I)

We can express the field to the right of the block (at $z = \Delta_{02}$), in terms of the phase transformation introduced by the glass, as

$$\psi_p(x, y, \Delta_{02}) = \exp\{-j[nk_0\Delta_2(x, y) + k_0[\Delta_{02} - \Delta_2(x, y)]]\}$$

$$(4.1\text{-}2)$$

Therefore, at a distance z' further away from the plane $z = \Delta_{02}$, we can find the field by convolving $\psi_p(x, y, \Delta_{02})$ with the spatial impulse response of propagation [see Eqs. (3.3-19a) and (3.3-19b)],

$$\psi_p(x, y, z' + \Delta_{02})$$

$$= \frac{jk_0}{2\pi z'}\exp(-jk_0 z')\iint\limits_{-\infty}^{\infty}\exp\{-j[nk_0\Delta_2(x', y')$$

$$+ k_0[\Delta_{02} - \Delta_2(x', y')]]\}$$

$$\times \exp\left\{-\frac{jk_0}{2z'}[(x - x')^2 + (y - y')^2]\right\} dx'\, dy'$$

$$= \frac{jk_0}{2\pi z'}\exp(-jk_0 z')\exp(-jk_0\Delta_{02})\exp\left[-j\frac{k_0}{2z'}(x^2 + y^2)\right]$$

$$\times \iint\limits_{-\infty}^{\infty}\exp[-jk_0(n - 1)\Delta_2(x', y')]\exp\left[j\frac{k_0}{z'}(xx' + yy')\right]$$

$$\times \exp\left[-j\frac{k_0}{2z'}(x'^2 + y'^2)\right] dx'\, dy'. \qquad (4.1\text{-}3)$$

In writing this equation, we have assumed the slab of glass to be infinitely wide. The effect of a finite aperture will be discussed later.

In Eq. (4.1-3), the last exponent within the integral indicates a quadratic phase curvature. In the derivation of the Fraunhofer diffraction formula from the Fresnel diffraction formula (see Section 3.5), a similar term was deleted on the basis of Eq. (3.5-1), the far-field approximation. However, note that there now exists a possibility, in Eq. (4.1-3), to balance out this term through a proper choice of $\Delta_2(x', y')$. In fact, the choice

$$\Delta_2(x', y') = a + \frac{b}{2}(x'^2 + y'^2) \qquad (4.1\text{-}4)$$

shows that the x', y'-dependent terms in the arguments of the first and third components in the integrand in Eq. (4.1-3) balance out at

$$z' = -\frac{(1/b)}{n-1} = f. \qquad (4.1\text{-}5)$$

The relevance of including a constant a in Eq. (4.1-4) will become clear shortly.

At $z' = f$ or $z = f + \Delta_{02}$, the field is proportional to

$$\int\!\!\int_{-\infty}^{\infty} \exp\left[j\frac{k_0}{f}(xx' + yy') \right] dx'\, dy' = \mathscr{F}_{xy}\{1\}\Big|_{\substack{k_x = k_0 x/f, \\ k_y = k_0 y/f}} \qquad (4.1\text{-}6)$$

which is proportional to a δ function in x and y. We call the distance $z' = f$ the *focal length*, because the plane wave incident on the left of the glass has been brought to a point, or *focus*, justifying the symbol f. We can also view the *focal plane* $z = f + \Delta_{02}$ as containing the Fourier transform of the amplitude distribution of the field incident on the glass slab.

Now recall from elementary analytical geometry that the *radius of curvature* R_2 of the function $\Delta_2(x', y')$ in Eq. (4.1-4) is approximately given by $1/b$ under the paraxial approximation $[b(x'^2 + y'^2) \ll 1]$. In Eq. (4.1-5), we can replace by $1/b$ by R_2 to get

$$f = -\frac{R_2}{n-1}. \qquad (4.1\text{-}7)$$

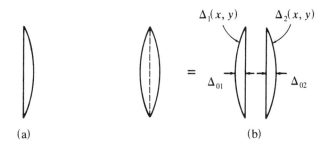

Figure 4.2 (a) A plano-convex lens. (b) A biconvex lens as the superposition of two plano-convex lenses back-to-back.

We now choose $a = \Delta_{02}$ to ensure maximum thickness of the slab of glass on-axis ($x', y' = 0$). Thus, Eq. (4.1-4) becomes

$$\Delta_2(x, y) = \Delta_{02} + \frac{x^2 + y^2}{2R_2}. \qquad (4.1\text{-}8)$$

For a positive value of f, R_2 must be negative. We remind readers of the sign convention in optics introduced in Chapter 2: As rays (or waves) travel from left to right, each concave surface is taken to have a negative radius of curvature, as shown in Figure 4.2(a) for a *plano-convex lens*. If, on the other hand, the rays see a convex surface, the surface has a positive radius of curvature.

Consider, now, two plano-convex lenses placed back-to-back, as shown in Figure 4.2(b). This constitutes a *biconvex* (or, simply, *convex*) lens. We can express the thickness function $\Delta(x, y)$ for a biconvex lens in terms of the thickness functions $\Delta_1(x, y)$ and $\Delta_2(x,y)$ of the plano-convex lenses, according to

$$\Delta(x, y) = \Delta_1(x, y) + \Delta_2(x, y)$$

$$= \left[\Delta_{01} - \frac{1}{2R_1}(x^2 + y^2)\right] + \left[\Delta_{02} + \frac{1}{2R_2}(x^2 + y^2)\right]$$

$$= \Delta_0 - \frac{1}{2}(x^2 + y^2)\left(\frac{1}{R_1} - \frac{1}{R_2}\right). \qquad (4.1\text{-}9)$$

In Eq. (4.1-9), Δ_{01} and Δ_{02} represent the peak thicknesses of the plano-convex lenses, and R_1 (> 0) and R_2 (< 0) are their radii of curvatures, respectively. The effective focal length of the (bi)convex lens can then be derived as

$$\frac{1}{f} = (n - 1)\left(\frac{1}{R_1} - \frac{1}{R_2}\right). \qquad (4.1\text{-}10)$$

Note that this expression for the focal length is identical to that derived from ray optics [see Eq. (2.4-17)]. Equation (4.1-10) holds true for concave lenses as well.

Now, we can express the phase transformation introduced by the biconvex lens as [see Eq. (4.1-2)]

$$\exp\left[-j\left(nk_0\Delta(x, y) + k_0(\Delta_0 - \Delta(x, y))\right)\right]$$

$$= \exp\left[-jk_0\Delta_0\right]\exp\left[-j(n - 1)k_0\Delta(x, y)\right].$$

Substituting for $\Delta(x, y)$ from Eq. (4.1-9) and using Eq. (4.1-10), the preceding expression becomes equal to $\exp[-jnk_0\Delta_0]\exp[j(k_0/2f)(x^2 + y^2)]$. Then for a field ψ_{pL} incident on the lens, the field ψ_{pR} immediately behind it is

$$\psi_{pR} = \psi_{pL}\exp(-jnk_0\Delta_0)\exp\left[j\frac{k_0}{2f}(x^2 + y^2)\right]. \qquad (4.1\text{-}11a)$$

$$\simeq \psi_{pL}\exp\left[j\frac{k_0}{2f}(x^2 + y^2)\right], \qquad (4.1\text{-}11b)$$

if the lens is a *thin lens* (i.e., $k_0\Delta_0 \ll 1$). We will assume all lenses to be thin lenses in the remainder of this text.

Let us now investigate the effect of placing a *transparency* $t(x, y)$ immediately against a thin lens, whether in front or behind, as shown in Figure 4.3. In general, $t(x, y)$ a complex function such that if a complex field $\psi_p(x, y)$ is incident on it, the field *immediately behind* it is given by $\psi_p(x, y)t(x, y)$, where we have assumed that the transparency is infinitely thin. Then, under illumination by a unit-amplitude plane wave, the field immediately behind the

Optical Information Processing (I)

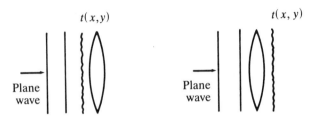

Figure 4.3 A transparency immediately before and after a lens under plane-wave illumination.

transparency–lens combination is given by

$$t(x, y)\exp\left[j\frac{k_0}{2f}(x^2 + y^2)\right].$$

We then find the field at a distance $z \simeq z' = f$ by using the Fresnel diffraction formula, Eq. (3.3-19), as

$$\psi_p|_{z=f} \triangleq \psi_p(x, y; f)$$

$$= \frac{jk_0}{2\pi f}\exp(-jk_0 f)\exp\left[-j\frac{k_0}{2f}(x^2 + y^2)\right]$$

$$\times \int\!\!\!\int_{-\infty}^{\infty} t(x', y')\exp\left[j\frac{k_0}{f}(xx' + yy')\right] dx'\, dy' \qquad (4.1\text{-}12a)$$

$$= \frac{jk_0}{2\pi f}\exp(-jk_0 f)\exp\left[-j\frac{k_0}{2f}(x^2 + y^2)\right]$$

$$\times \mathcal{F}_{xy}[t(x, y)]\Big|_{\substack{k_x=k_0 x/f, \\ k_y=k_0 y/f}} \qquad (4.1\text{-}12b)$$

where x and y denote the transverse coordinates at $z = f$. Hence, the complex field on the focal plane ($z = f$) is proportional to the Fourier transform of $t(x, y)$ but with a phase curvature [depicted by the second exponential on the RHS of Eq. (4.1-12a)].

We remark that if the transparency is placed immediately behind the lens, the field illuminating the transparency for a plane wave incident on the lens is given by Eq. (4.1-11) as proportional to

$$\exp\left[j\frac{k_0}{2f}(x^2 + y^2)\right],$$

which defines a converging beam. This proves that the advantage of illuminating a transparency by a converging beam instead of a parallel beam (as in the case of plane-wave illumination) lies in bringing the transform plane from the far field (as in Fraunhofer diffraction) to a finite distance away, that is, the plane of the focus of the converging lens, as previously illustrated also by Problem 3–10.

However, all physical lenses have finite apertures. We can model this physical situation as a lens with an infinite aperture followed immediately by a transparency described by what is called the *pupil function* $p_f(x, y)$ of the lens. Typical pupil functions are $\text{rect}(x/X)\text{rect}(y/Y)$ or $\text{circ}(r/R)$, where X, Y, and R are some constants. Hence, if we have a transparency $t(x, y)$ against a lens with a finite aperture, the field at the back focal plane of the lens is given by

$$\psi_p(x, y; f) \propto \mathscr{F}_{xy}\{t(x, y)p_f(x, y)\}\Big|_{\substack{k_x = k_0 x/f \\ k_y = k_0 y/f}}$$

$$= \frac{1}{4\pi^2}\left(\mathscr{F}_{xy}\{t(x, y)\} * \mathscr{F}_{xy}\{p_f(x, y)\}\right)\Big|_{\substack{k_x = k_0 x/f \\ k_y = k_0 y/f}} \quad (4.1\text{-}13)$$

under plane-wave illumination.

Example 4.1 Transparency in Front of a Lens

Suppose that a transparency $t(x, y)$ is located at a distance d_o in front of a convex lens with an infinitely large aperture and is illuminated by a plane wave (see Figure 4.4). Assuming the field to the left of $t(x, y)$ to be of unit strength, the field to the right of the transparency is $t(x, y)$. This travels a distance d_o to the lens; hence, using the transfer function approach to wave propagation,

Optical Information Processing (I)

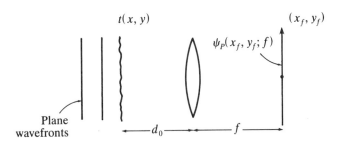

Figure 4.4 Plane-wave illumination of a transparency $t(x, y)$ located a distance d_o in front of a converging lens of focal length f.

and Eq. (3.3-14), we have that

$$\Psi_e(k_x, k_y)\Big|_{\substack{\text{front} \\ \text{of lens}}} = T(k_x, k_y)\exp\left[\frac{j(k_x^2 + k_y^2)d_o}{2k_0}\right], \quad (4.1\text{-}14)$$

where

$$T(k_x, k_y) = \mathscr{F}_{xy}\{t(x, y)\}$$

and

$$\Psi_e(k_x, k_y)\Big|_{\substack{\text{front} \\ \text{of lens}}} = \mathscr{F}_{xy}\{\psi_e(x, y)\Big|_{\substack{\text{front} \\ \text{of lens}}}\}.$$

In what follows, we will state the method that we use to find the field in the back focal plane of the lens. We then write down the final expression in terms of Fourier transform operators, in order to avoid repeating lengthy but similar mathematical expressions. First, note that by taking the inverse Fourier transform of Eq. (4.1-14) we get $\psi_e(x, y)$, which when multiplied by $e^{-jk_0 d_o}$, yields $\psi_p(x, y)$ immediately in front of the lens, in accordance with Eq. (1.4-4). Hence, the complex amplitude in the back focal plane of the lens can be found from Eq. (4.1-12b) by replacing $t(x, y)$ by the field immediately in front of the lens. This gives

$$\psi_p(x, y; f) = \frac{jk_0}{2\pi f} \exp(-jk_0 d_o) \exp(-jk_0 f)$$

$$\times \exp\left[-j\frac{k_0}{2f}(x^2 + y^2)\right]$$

$$\times \mathscr{F}_{xy}\left\{\mathscr{F}_{xy}^{-1}\left\{T(k_x, k_y)\exp\left[\frac{j(k_x^2 + k_y^2)d_o}{2k_0}\right]\right\}\right\}\Bigg|_{\substack{k_x = k_0 x/f \\ k_y = k_0 y/f}}$$

$$= \frac{jk_0}{2\pi f}\exp\left[-jk_0(d_o + f)\right]$$

$$\times \exp\left[-j\frac{k_0}{2f}\left(1 - \frac{d_o}{f}\right)(x^2 + y^2)\right]\{T(k_x, k_y)\}\Bigg|_{\substack{k_x = k_0 x/f \\ k_y = k_0 y/f}}$$

$$(4.1\text{-}15)$$

Note that, as in Eq. (4.1-12), a phase curvature factor again pre-cedes the Fourier transform, but vanishes for the special case $d_o = f$. Thus, *when the object (transparency) is placed in the front focal plane of the convex lens, the phase curvature disappears, and we recover the exact Fourier transform on the back focal plane.* Fourier processing on an "input" transparency located on the front focal plane may now be performed on the back focal plane, as will be seen later on in this chapter.

4.2 Imaging by a Single Lens and the Impulse Response of the Imaging System

So far, we have only concentrated on the nature of the field distribution at the back focal plane of the lens. Let us now examine the field distribution at an arbitrary distance behind the lens when a transparency, placed a distance d_o in front of the lens, is illumi-nated by a plane wave. Our motive is to verify the imaging property

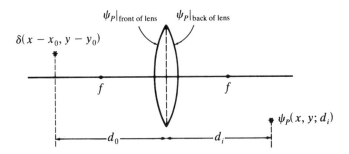

Figure 4.5 Single-lens imaging system with a point object.

of lenses, discussed earlier in Chapter 2. For simplicity, we first consider a point source as our object or input to the optical system. Mathematically, we can represent the object or the input transparency as $\delta(x - x_o, y - y_o)$, as shown in Figure 4.5. Now, we can track the propagation of the complex field from the input (or object) plane to the plane of the lens by using either the Fresnel diffraction formula or the transfer function approach to wave propagation. Using the former, we have, from Eq. (3.3-19),

$$\psi_p\big|_{\substack{\text{front} \\ \text{of lens}}} = \frac{jk_0}{2\pi d_o} \exp(-jk_0 d_o)\exp\left\{-j\frac{k_0}{2d_o}\left[(x - x_o)^2 + (y - y_o)^2\right]\right\}.$$

$$(4.2\text{-}1)$$

Alternatively, Eq. (4.2-1) can be directly written by realizing that we have a point source on the object plane. The field immediately behind the lens is given by multiplying Eq. (4.2-1) by the phase transformation introduced by the lens [Eq. (4.1-11b)]:

$$\psi_p\big|_{\substack{\text{back} \\ \text{of lens}}} = \frac{jk_0}{2\pi d_o} \exp(-jk_0 d_o)\exp\left[j\frac{k_0}{2f}(x^2 + y^2)\right]$$

$$\times \exp\left\{-j\frac{k_0}{2d_o}\left[(x - x_o)^2 + (y - y_o)^2\right]\right\}. \quad (4.2\text{-}2)$$

Finally, we can express the field at an arbitrary distance z to the right of the lens by the Fresnel diffraction formula [Eq. (3.3-19)] as

$$\psi_p(x, y, z) = \left(\frac{jk_0}{2\pi d_o}\right)\left(\frac{jk_0}{2\pi z}\right)\exp(-jk_0 d_o)\exp(-jk_0 z)$$

$$\times \int\int_{-\infty}^{\infty}\exp\left[j\frac{k_0}{2f}(x'^2 + y'^2)\right]$$

$$\times \exp\left\{-j\frac{k_0}{2d_o}\left[(x' - x_o)^2 + (y' - y_o)^2\right]\right\}$$

$$\times \exp\left\{-j\frac{k_0}{2z}\left[(x - x')^2 + (y - y')^2\right]\right\} dx'\, dy'$$

$$= -\frac{k_0^2}{4\pi^2 d_o z}\exp[-jk_0(d_o + z)]$$

$$\times \exp\left[-j\frac{k_0}{2d_o}(x_o^2 + y_o^2)\right]\exp\left[-j\frac{k_0}{2z}(x^2 + y^2)\right]$$

$$\times \int\int_{-\infty}^{\infty}\exp\left\{-j\frac{k_0}{2}\left[\frac{1}{d_o} + \frac{1}{z} - \frac{1}{f}\right](x'^2 + y'^2)\right\}$$

$$\times \exp\left\{jk_0\left[\left(\frac{x_o}{d_o} + \frac{x}{z}\right)x' + \left(\frac{y_o}{d_o} + \frac{y}{z}\right)y'\right]\right\} dx'\, dy'.$$

$$(4.2\text{-}3)$$

In Eq. (4.2-3), note that at a distance $z = d_i$ such that

$$\frac{1}{d_o} + \frac{1}{d_i} = \frac{1}{f}, \qquad (4.2\text{-}4)$$

the argument of the first exponential under the integral vanishes.

The integral now reduces to

$$
\mathscr{I} = \int\int_{-\infty}^{\infty} \exp\left\{ jk_0\left[\left(\frac{x_o}{d_o} + \frac{x}{d_i}\right)x' + \left(\frac{y_o}{d_o} + \frac{y}{d_i}\right)y'\right]\right\} dx'\, dy'
$$

$$
= 4\pi^2\delta\left(k_0\left(\frac{x_o}{d_o} + \frac{x}{d_i}\right), k_0\left(\frac{y_o}{d_o} + \frac{y}{d_i}\right)\right)
$$

$$
= 4\pi^2\delta\left(\frac{k_0}{d_i}(x - Mx_o), \frac{k_0}{d_i}(y - My_o)\right) \qquad \left(M \triangleq -\frac{d_i}{d_o}\right) \quad (4.2\text{-}5a)
$$

$$
= 4\pi^2\frac{d_i^2}{k_0^2}\delta(x - Mx_o, y - My_o), \qquad\qquad\qquad\qquad (4.2\text{-}5b)
$$

Incorporating all this into Eq. (4.2-3),

$$
\psi_p|_{z=d_i} \triangleq \psi_{pi} = M\exp[-jk_0(d_o + d_i)]\exp\left[-j\frac{k_0}{2d_o}(x_o^2 + y_o^2)\right]
$$

$$
\times \exp\left[-j\frac{k_0}{2d_i}(x^2 + y^2)\right]\delta(x - Mx_o, y - My_o).
$$

$$
(4.2\text{-}6)
$$

Thus, at $z = d_i$, the field is a δ function located at $x = Mx_o = -(d_i/d_o)x_o$, $y = My_o = -(d_i/d_o)y_o$. Hence we have *imaged* a point located at $(x = x_o, y = y_o)$ on the object plane (a distance d_o in front of the lens) at a point $(x = Mx_o, y = My_o)$ on the *image plane* (a distance d_i behind the lens). Equation (4.2-4) is in agreement with Eq. (2.4-19) derived using ray optics in Chapter 2. Recall that M is the magnification factor. Equation (4.2-6) is the impulse response of the single-lens *imaging system* due to an offset δ function in the object plane.

It is not hard to realize that because $\delta(x - x', y - y')$ on the object plane yields a response proportional to $M\delta(x - Mx', y - My')$ on the image plane according to the correspondence

$$
\delta(x - x', y - y') \rightarrow M\delta(x - Mx', y - My'),
$$

then

$$t(x', y')\delta(x - x', y - y') \rightarrow Mt(x', y')\delta(x - Mx', y - My'),$$

so that

$$t(x, y) = \int\limits_{-\infty}^{\infty}\!\!\!\int t(x', y')\delta(x - x', y - y')\, dx'\, dy'$$

$$\rightarrow M \int\limits_{-\infty}^{\infty}\!\!\!\int t(x', y')\delta(x - Mx', y - My')\, dx'\, dy'$$

$$= \frac{1}{M} \int\limits_{-\infty}^{\infty}\!\!\!\int t(x', y')\delta\!\left(x' - \frac{x}{M}, y' - \frac{y}{M}\right) dx'\, dy'$$

$$= \frac{1}{M} t\!\left(\frac{x}{M}, \frac{y}{M}\right). \tag{4.2-7}$$

Equation (4.2-7) tells us that a transparency $t(x, y)$ placed on the object plane produces a field on the image plane proportional to $(1/M)t(x/M, y/M)$ which is called the *geometrical image* of the object. Note that the size of the image is magnified by a factor $|M|$ and that a real image is inverted with respect to the object (because $M = -d_i/d_o < 0$), as expected from ray optics. Observe also that the field amplitude is reduced by a factor M.

   Note that in the preceding discussion, we avoided a serious complication by tacitly stating that an input $\delta(x - x', y - y')$ on the object plane yields a response proportional to $M\delta(x - Mx', y - My')$ on the image plane. Strictly speaking, however, the complex field on the image plane also has three exponential terms as shown in Eq. (4.2-6). Although the first one poses no problem, the second and third represent quadratic phase curvatures. If we were willing to consider image formation between two spherical surfaces, rather than between two planes, then we can easily show that the two terms have no effect on imaging. We will sketch here an argument to prove that there is minimal effect from these terms *even if we consider image formation between two planes*. Note that, in reality, the field amplitude at a point (x, y) on the image plane consists of contributions only from a tiny region of object space, namely, around the corresponding point (x_o, y_o). Also within this region, the argument of the second exponential in Eq. (4.2-6) hardly changes. Hence, the second and third exponents may be approximately combined as

$$\exp\left[-j\frac{k_0}{2}\left(\frac{1}{M^2 d_o} + \frac{1}{d_i}\right)(x^2 + y^2)\right]$$

$$= \exp\left[-j\frac{k_0}{2d_i}\left(1 - \frac{1}{M}\right)(x^2 + y^2)\right].$$

Now, because in a majority of cases, we will only be interested in the distribution of *light intensity* on the image plane of the lens, the exponents will not create any major problem and can therefore be dropped. For a more comprehensive argument, we refer the readers to Goodman (1968).

With potential complications put to rest once and for all, we return to the mainstream of things and consider a single-lens imaging system as shown in Figure 4.5, but where the lens has a pupil function $p_f(x, y)$ placed against the lens. Note that if $p_f(x, y) = 1$, the response of the imaging system is identical to the geometrical image of the object [see Eq. (4.2-7)]. However, for a general pupil function, this is not the case. Our objective is now to characterize the imaging system in terms of its transfer function and impulse response, similar in idea to the transfer function and impulse response of propagation that was discussed in the previous chapter.

In attempting to find the impulse response of this system, we consider, as before, the object to be a point source $\delta(x - x_o, y - y_o)$, so that the complex field in front of the lens is once again given by Eq. (4.2-1). The field immediately behind the lens is similar to Eq. (4.2-2), with the RHS multiplied by $p_f(x, y)$. We can then express the field ψ_{pi} at a distance $z = d_i$ behind the lens by using the Fresnel diffraction formula. The result is

$$\psi_{pi}(x, y) = -\frac{k_0^2}{4\pi^2 d_o d_i}\exp[-jk_0(d_o + d_i)]$$

$$\times \exp\left[-j\frac{k_0}{2d_o}(x_o^2 + y_o^2)\right]\exp\left[-j\frac{k_0}{2d_i}(x^2 + y^2)\right]\mathscr{I},$$

$$(4.2\text{-}8)$$

where the integral $\mathscr{I} = \mathscr{I}(x, y)$ is given as

$$\mathscr{I} = \int\!\!\!\int_{-\infty}^{\infty} p_f(x', y') \exp\left\{ jk_0\left[\left(\frac{x_o}{d_o} + \frac{x}{d_i}\right)x' + \left(\frac{y_o}{d_o} + \frac{y}{d_i}\right)y' \right]\right\} dx'\, dy'$$

$$= P_f\left(k_0\left(\frac{x_o}{d_o} + \frac{x}{d_i}\right), k_0\left(\frac{y_o}{d_o} + \frac{y}{d_i}\right)\right), \qquad (4.2\text{-}9a)$$

where

$$P_f(k_x, k_y) \triangleq \mathscr{F}_{xy}\left[p_f(x, y)\right]. \qquad (4.2\text{-}9b)$$

Note that for $p_f(x, y) = 1$, Eq. (4.2-9a) reduces to the expression for $\mathscr{I}$ in Eq. (4.2-5), which is the case for imaging with a lens having an infinite aperture.

From Eqs. (4.2-8) and (4.2-9), we can write

$$\psi_{pi} = -\frac{k_0^2}{4\pi^2 d_o d_i} P_f\left(k_0\left(\frac{x_o}{d_o} + \frac{x}{d_i}\right), k_0\left(\frac{y_o}{d_o} + \frac{y}{d_i}\right)\right). \quad (4.2\text{-}10)$$

In writing Eq. (4.2-10), the complex term due to propagation, namely, $\exp[-jk_0(d_o + d_i)]$, has been disregarded, and the quadratic phase curvature terms, namely, $\exp[-j(k_0/2d_o)(x_o^2 + y_o^2)]$ and $\exp[-j(k_0/2d_i)(x^2 + y^2)]$ have been dropped on the basis of the arguments advanced previously. Strictly speaking, the contributions arising from these phase curvature terms contaminate the exact Fourier transform of the pupil function $p_f(x, y)$ given by Eq. (4.2-10). However, such phase distortions do not appear in the two-lens system considered in the next section.

Having determined the impulse response of a single-lens imaging system where the lens has a pupil function $p_f(x, y)$, we now find the response of the system for an arbitrary input $t(x, y)$. To do this, note that $\delta(x - x', y - y')$ on the object plane yields a response according to [see Eq. (4.2-10)]

$$\delta(x - x', y - y') \rightarrow -\frac{k_0^2}{4\pi^2 d_o d_i} P_f\left(k_0\left(\frac{x'}{d_o} + \frac{x}{d_i}\right), k_0\left(\frac{y'}{d_o} + \frac{y}{d_i}\right)\right)$$

Optical Information Processing (I)

on the image plane; hence,

$$t(x, y) = \iint\limits_{-\infty}^{\infty} t(x', y')\delta(x - x', y - y')\, dx'\, dy'$$

$$\rightarrow -\frac{k_0^2}{4\pi^2 d_o d_i} \iint\limits_{-\infty}^{\infty} t(x', y') P_f\left(k_0\left(\frac{x'}{d_o} + \frac{x}{d_i}\right), k_0\left(\frac{y'}{d_o} + \frac{y}{d_i}\right)\right)$$

$$\times dx'\, dy' \tag{4.2-11}$$

$$= -\frac{k_0^2}{4\pi^2 d_o d_i M^2} \iint\limits_{-\infty}^{\infty} t\left(\frac{x''}{M}, \frac{y''}{M}\right) P_f\left(\frac{k_0}{d_i}(x - x''), \frac{k_0}{d_i}(y - y'')\right)$$

$$\times dx''\, dy'' \left(\text{with } x' = \frac{x''}{M},\ y' = \frac{y''}{M}\right)$$

$$= -\frac{k_0^2}{4\pi^2 d_o d_i M^2}\left(t\left(\frac{x}{M}, \frac{y}{M}\right) * P_f\left(\frac{k_0}{d_i}x, \frac{k_0}{d_i}y\right)\right). \tag{4.2-12}$$

The result [Eq. (4.2-12)] indicates that a transparency $t(x, y)$ placed on the object plane produces a field on the image plane proportional to the geometrical image $t(x/M, y/M)$ convolved with the *impulse response* of the imaging system defined as

$$\mathcal{k}(x, y) = P_f\left(\frac{k_0}{d_i}x, \frac{k_0}{d_i}y\right). \tag{4.2-13}$$

The Fourier transform of $\mathcal{k}(x, y)$ will be called the *coherent transfer function*, given by

$$\mathcal{H}(k_x, k_y) = p_f\left(-\frac{d_i}{k_0}k_x, -\frac{d_i}{k_0}k_y\right). \tag{4.2-14}$$

In writing Eq. (4.2-14), we have neglected a multiplicative constant.
We remark that, in retrospect, the impulse response in Eq. (4.2-13) can be obtained directly by setting $x_o = 0 = y_o$ in Eq. (4.2-10) and disregarding the constant factor on the RHS.

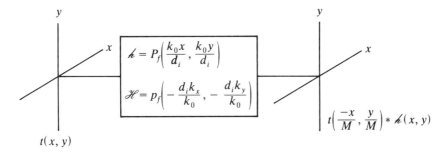

Figure 4.6 Block diagram representation of the impulse response and the coherent transfer function of an imaging system.

The concepts just advanced are summarized in Figure 4.6. We end this section with a reminder that the pupil function p_f can represent, in effect, an arbitrary transparency placed immediately against the lens. The field on the image plane is

$$\psi_{pi}(x, y) \propto t\left(\frac{x}{M}, \frac{y}{M}\right) * \hbar(x, y) \qquad (4.2\text{-}15)$$

and hence the corresponding image intensity is

$$I_i(x, y) = |\psi_{pi}|^2 \propto \left| t\left(\frac{x}{M}, \frac{y}{M}\right) * \hbar(x, y) \right|^2. \qquad (4.2\text{-}16)$$

An example on the use of the preceding equations follows later, in Section 4.5.

4.3 Impulse Response of a Two-Lens System

While on the topic of impulse response and the coherent transfer function, it is instructive to discuss these with reference to a two-lens system as shown in Figure 4.7. The two-lens system is traditionally attractive for image processing purposes because, in the configuration shown in the figure, the Fourier transform of the input transparency appears on the common focal plane, or *Fourier plane*. In order to perform Fourier-plane processing on the input transparency, we can insert a transparency on the Fourier plane that will suitably modify the Fourier transform of the input transparency. This Fourier plane transparency is commonly called a

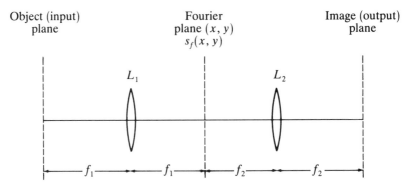

Figure 4.7 A two-lens image processing system.

spatial filter $s_f(x, y)$. Examples of image processing using a two-lens system appear in Section 4.5.

In order to find the impulse response of this two-lens image processing system, we once again put a point object $\delta(x - x_o, y - y_0)$ on the input or object plane. For simplicity, we assume that the lenses L_1 and L_2 have infinitely wide apertures (or pupils). We can find the complex field on the back focal plane of L_1 and immediately in front of the spatial filter $s_f(x, y)$ by using Eq. (4.1-15) with $f = d_o = f_1$. Neglecting the constant phase term in Eq. (4.1-15), we then have

$$\psi_p(x, y) = \frac{jk_0}{2\pi f_1} \exp\left[\frac{jk_0(x_o x + y_o y)}{f_1}\right]$$

immediately in front of the spatial filter on the Fourier plane. The field immediately behind the spatial filter is simply the preceding expression multiplied by $s_f(x, y)$. We use Eq. (4.1-15) again to find the field on the back focal plane (or image plane) of the second lens L_2. Straightforward calculations show that this results in

$$\psi_{pi}(x, y) \propto -\frac{k_0^2}{4\pi^2 f_1 f_2} \mathscr{F}_{xy}\left(\exp\left[\frac{jk_0(x_o x + y_o y)}{f_1}\right] s_f(x, y)\right)\bigg|_{\substack{k_x = k_0 x/f_2 \\ k_y = k_0 y/f_2}}$$

$$= -\frac{k_0^2}{4\pi^2 f_1 f_2} S_f\left(k_0\left(\frac{x_o}{f_1} + \frac{x}{f_2}\right), k_0\left(\frac{y_o}{f_1} + \frac{y}{f_2}\right)\right), \quad (4.3\text{-}1)$$

where

$$S_f(k_x, k_y) \triangleq \mathscr{F}_{xy}\left[s_f(x, y)\right]. \qquad (4.3\text{-}2)$$

Comparing Eq. (4.3-1) with Eq. (4.2-10), which describes the impulse response of a one-lens system, we see the striking similarities: The spatial filter $s_f(x, y)$ in the two-lens system can be compared with the pupil function $p_f(x, y)$ of the single-lens system, and f_1 and f_2 take on the roles of d_o and d_i, respectively. We may go even a step further and observe that a two-lens system can be replaced by an equivalent single-lens system with the spatial filter s_f of the two-lens system now being placed against the lens in the single-lens system. The object and image distances of the equivalent one-lens system would have to be f_1 and f_2, respectively, and the focal length f_e of the equivalent lens could be found using the lens formula and reads

$$\frac{1}{f_1} + \frac{1}{f_2} = \frac{1}{f_e}. \qquad (4.3\text{-}3)$$

Astonishingly, this reads like the equivalent focal length of a two-lens combination placed back-to-back [see Eq. (2.4-23b)]! This amounts to saying that the two-lens combination shown in Figure 4.7 may be converted to an equivalent single-lens system by "collapsing" the two lenses into one, taking care to maintain the object and image distances at f_1 and f_2.

We caution readers that the preceding comparison is only approximate, because the impulse response of a single-lens system is contaminated by the presence of phase curvatures. This is why the two-lens system is often preferred over the single-lens system for image processing applications.

Returning to the mainstream of our discussion, we can define the impulse response of the two-lens system, as in the single-lens case, by setting $x_0 = 0 = y_0$ in Eq. (4.3-1) and neglecting the constant factor on the RHS:

$$h(x, y) = S_f\left(\frac{k_0 x}{f_2}, \frac{k_0 y}{f_2}\right). \qquad (4.3\text{-}4)$$

Optical Information Processing (I)

The corresponding coherent transfer function is

$$\mathcal{H}(k_x, k_y) = s_f\left(-\frac{f_2}{k_0}k_x, -\frac{f_2}{k_0}k_y\right). \qquad (4.3\text{-}5)$$

We can show that the field on the back focal plane of the second lens L_2 for an input transparency $t(x, y)$ at the front focal plane of L_1 is given by Eq. (4.2-15) and the image intensity by Eq. (4.2-16), where $k(x, y)$ is defined in Eq. (4.3-4) and with $M = -f_2/f_1$.

4.4 Fourier-Transforming and Imaging Properties of Spherical Mirrors

In our treatment of lenses thus far, it is important to realize that the role of a lens in Fourier-transforming and image formation is primarily to introduce a phase curvature of the form $\exp[j(k_0/2f)(x^2 + y^2)]$. The same effect can be introduced by spherical mirrors. For instance, when a plane wave strikes a concave mirror, the reflected wavefronts are curved and converge to the focus. The only difference is that in mirrors the phase curvatures are introduced through reflection rather than transmission as in the case of lenses. Thus, if we take d_o and d_i to be the object and image distances during imaging by a mirror of radius R, where d_o and d_i are now on the same side of the mirror, the theory of image formation is exactly the same as before, and the relation in Eq. (4.2-4) may be readily rederived with $f = -R/2$. Also, on the focal plane, the reflected field will be proportional to the Fourier transform of the object, as described by Eq. (4.1-15).

4.5 Imaging with Spatially Incoherent Light and the Optical Transfer Function

Our discussion on diffraction and the Fourier-transforming and imaging properties of lenses has so far been based on the fact that the illumination of the object is *spatially coherent*. This means that the complex amplitudes of light falling on all parts of the object vary in unison, that is, any two points on the object receive light that has a fixed relative phase that does not vary with time. Before the advent of lasers, which form excellent coherent sources (see Chapter 7), spatially coherent illumination could be simulated by starting from an intense point source of light and collimating the light by a lens.

Conversely, an object may be illuminated with light having the property that the phasor amplitudes on all parts of the object vary randomly, so that any two points on the object receive light that has a *random* relative phase that changes with time. This sort of illumination is termed *spatially incoherent*. Light from diffuse or extended sources, such as flourescent tube lights, is incoherent.

How does this affect our discussion on the impulse response and transfer function of imaging systems? Returning to Eq. (4.2-15), we recast it in the form

$$\psi_{pi}(x, y) \propto \psi_{p0}\left(\frac{x}{M}, \frac{y}{M}\right) * \mathscr{k}(x, y)$$

$$= \iint_{-\infty}^{\infty} \psi_{p0}\left(\frac{x'}{M}, \frac{y'}{M}\right) \mathscr{k}(x - x', y - y')\, dx'\, dy'. \quad (4.5\text{-}1)$$

In the preceding equation, we have written $\psi_{p0}(x, y)$ instead of $t(x,y)$, where ψ_{p0} denotes the complex field on the object plane immediately behind the transparency $t(x, y)$, resulting from the illumination of the transparency by an arbitrary field distribution. Equation (4.5-1) holds true for both one- and two-lens imaging systems. Now, we can express the image intensity $I_i = |\psi_{pi}|^2$ as

$$I_i(x, y) \propto \iiiint_{-\infty}^{\infty} \mathscr{k}(x - x', y - y')\mathscr{k}^*(x - x'', y - y'')$$

$$\times \left\langle \psi_{p0}\left(\frac{x'}{M}, \frac{y'}{M}; t\right)\psi_{p0}^*\left(\frac{x''}{M}, \frac{y''}{M}; t\right)\right\rangle dx'\, dy'\, dx''\, dy'', \quad (4.5\text{-}2)$$

where we have purposely introduced a time average over a temporal period of the optical field (denoted by $\langle \ \rangle$) and put in a time dependence in ψ_{p0} to accommodate possible spatial incoherence. The reason the time dependence is important is that, in order to observe whether illumination is spatially incoherent, we have to monitor two points on the object over a length of time.

In the case of coherent object illumination, the complex amplitudes across the object plane differ by complex constants instead of in a random fashion. It suffices, therefore, to consider $\psi_{p0}(x', y')$ and $\psi_{p0}(x'', y'')$ at $t = 0$. The time average in Eq. (4.5-2) drops out, yielding

$$I_i(x, y) \propto \left| \int\int_{-\infty}^{\infty} \textit{k}(x - x', y - y')\psi_{p0}\left(\frac{x'}{M}, \frac{y'}{M}\right) dx' \, dy' \right|^2 \quad (4.5\text{-}3a)$$

$$= \left| \textit{k}(x, y) * t\left(\frac{x}{M}, \frac{y}{M}\right) \right|^2. \quad (4.5\text{-}3b)$$

In going from Eq. (4.5-3a) to (4.5-3b), we have assumed plane-wave illumination of the transparency $t(x, y)$. Equation (4.5-3b) is identical to Eq. (4.2-16).

For the case of incoherent object illumination, we make use of the result [see Beran and Parrent (1963)]

$$\left\langle \psi_{p0}\left(\frac{x'}{M}, \frac{y'}{M}; t\right)\psi_{p0}^*\left(\frac{x''}{M}, \frac{y''}{M}; \right) \right\rangle \propto \left| \psi_{p0}\left(\frac{x'}{M}, \frac{y'}{M}\right) \right|^2 \delta(x' - x'', y' - y'')$$

$$(4.5\text{-}4)$$

to recast Eq. (4.5-2) in the form

$$I_i(x, y) \propto \int\int_{-\infty}^{\infty} |\textit{k}(x - x', y - y')|^2 \left| \psi_{p0}\left(\frac{x'}{M}, \frac{y'}{M}\right) \right|^2 dx' \, dy', \quad (4.5\text{-}5a)$$

or, equivalently, as

$$I_i(x, y) \propto |\textit{k}(x, y)|^2 * \left| t\left(\frac{x}{M}, \frac{y}{M}\right) \right|^2, \quad (4.5\text{-}5b)$$

assuming plane-wave illumination of the object transparency.

Equations (4.5-3) and (4.5-5) hold for both one- and two-lens systems.

We can summarize the essence of Eqs. (4.5-3) and (4.5-5) as follows: *A coherent system is linear with respect to the complex fields.*

On the other hand, *an incoherent system is linear with respect to the intensities.*

We remind readers that whereas the coherent transfer function $\mathscr{H}(k_x, k_y)$ relates the Fourier transform of the object field under coherent illumination to that of the image field, according to

$$\mathscr{F}_{xy}\left[\psi_{pi}(x, y)\right] \propto \mathscr{F}_{xy}\left[t\left(\frac{x}{M}, \frac{y}{M}\right)\right]\mathscr{H}(k_x, k_y) \qquad (4.5\text{-}6)$$

[see Eqs. (4.2-15)], the Fourier transform of the image intensity is related to the object field intensity under incoherent illumination according to

$$\mathscr{F}_{xy}\left[I_i(x, y)\right] \propto \mathscr{F}_{xy}\left[\left|t\left(\frac{x}{M}, \frac{y}{M}\right)\right|^2\right]\text{OTF}(k_x, k_y), \qquad (4.5\text{-}7)$$

where

$$\text{OTF}(k_x, k_y) \triangleq \mathscr{H}(k_x, k_y) \circledast \mathscr{H}(k_x, k_y) \qquad (4.5\text{-}8a)$$

is the *optical transfer function* of the incoherent imaging system. In Eq. (4.5-8a), the $\circledast$ denotes correlation. The OTF can be expressed explicitly as

$$\text{OTF}(k_x, k_y) \triangleq \int\!\!\!\int_{-\infty}^{\infty} \mathscr{H}^*(k_x', k_y')\mathscr{H}(k_x' + k_x, k_y' + k_y)\, dk_x'\, dk_y'$$

$$(4.5\text{-}8b)$$

$$= \int\!\!\!\int_{-\infty}^{\infty} \mathscr{H}^*\left(k_x' - \frac{k_x}{2}, k_y' - \frac{k_y}{2}\right)$$

$$\times \mathscr{H}\left(k_x' + \frac{k_x}{2}, k_y' + \frac{k_y}{2}\right) dk_x'\, dk_y'. \qquad (4.5\text{-}8c)$$

The relations in Eqs. (4.5-8a)–(4.5-8c) can be readily proved by formally writing down the Fourier transform of $|\hbar(x, y)|^2$ in terms of $\mathscr{H}(k_x, k_y)$; this is left as an exercise for the reader.

Optical Information Processing (I)

Some general properties of the OTF follow from the properties of correlation:

$$|\text{OTF}(k_x, k_y)| \leq |\text{OTF}(0,0)|; \qquad (4.5\text{-}9)$$

$$\text{OTF}(-k_x, -k_y) = \text{OTF}^*(k_x, k_y). \qquad (4.5\text{-}10)$$

In what follows, we will provide examples of OTFs and compare coherent and incoherent image processing.

Example 4.2 OTF of a Two-Lens System

Consider a two-lens system as shown in Figure 4.7, with $f_1 = f_2 = f$ and $s_f(x, y) = \text{rect}(x/X)\,\text{rect}(y/Y)$. Using Eq. (4.3-5), the coherent transfer function is

$$\mathscr{H}(k_x, k_y) = \text{rect}\left(\frac{k_x}{Xk_0/f}\right)\text{rect}\left(\frac{k_y}{Yk_0/f}\right). \qquad (4.5\text{-}11)$$

We plot $\mathscr{H}(k_x, k_y)$ versus k_x with k_y constant in Figure 4.8(a); the plot versus k_y can be drawn similarly. The OTF is the autocorrelation of $\mathscr{H}$ and is plotted in Figure 4.8(b). Observe that the *passband* of the OTF is twice as much as that of $\mathscr{H}$. This in turn implies that, using incoherent illumination, it is possible to transmit twice the range of spatial frequencies of the object than under

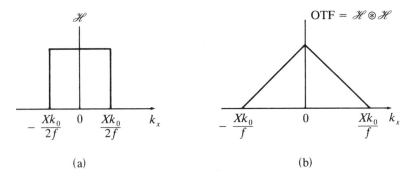

(a)   (b)

Figure 4.8 The coherent transfer function and the OTF of a two-lens system with $s_f(x, y) = \text{rect}(x/X)\text{rect}(y/Y)$.

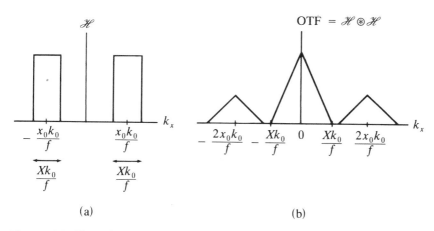

Figure 4.9 The coherent transfer function and the OTF of a two-lens system
with

$$s_f(x, y) = \left[\text{rect}\left(\frac{x - x_o}{X}\right) + \text{rect}\left(\frac{x + x_0}{X}\right)\right]\text{rect}\left(\frac{y}{Y}\right).$$

coherent illumination; however, the spectrum of the object transmitted through the passband is modified by the shape of the OTF.
Next, consider

$$s_f(x, y) = \left[\text{rect}\left(\frac{x - x_0}{X}\right) + \text{rect}\left(\frac{x + x_0}{X}\right)\right]\text{rect}\left(\frac{y}{Y}\right), \qquad x_0 > \frac{X}{2}.$$

We plot $\mathcal{H}(k_x, k_y)$ versus k_x with k_y constant in Figure 4.9(a)
along with the OTF in Figure 4.9(b). Note that even though it may
be positive to achieve band-pass characteristics with coherent illumination, incoherent processing always gives rise to inherently
low-pass characteristics. In the last decade, much attention has
been focussed on devising methods to realize band-pass characteristics using novel incoherent image processing techniques [see, e.g.,
Lohmann and Rhodes (1978), Stoner (1978), Poon and Korpel
(1979), and Glaser (1987)].

Example 4.3 Coherent versus Incoherent Imaging

Given a transparency $t(x, y) = \text{rect}(x/A)\text{rect}(y/A)$ and a
$1:1$ two-lens system using lenses of the same focal length f but
having $s_f(x, y) = \text{rect}(x/B)\text{rect}(y/B)$, the image intensities for

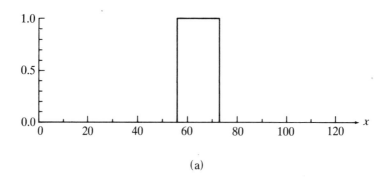

(a)

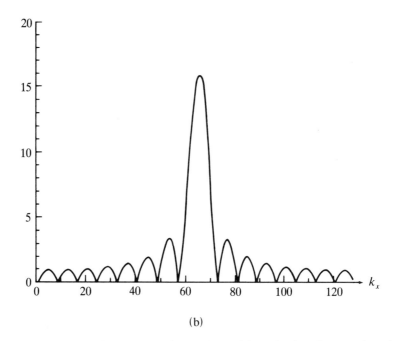

(b)

Figure 4.10 Calculation of the image intensities of $t(x, y) = \text{rect}(x/A) \times \text{rect}(y/A)$ formed by a two-lens imaging system with $s_f(x, y) = \text{rect}(x/B)\text{rect}(y/B)$ under coherent and incoherent illumination. For convenience, only x or k_x variations are drawn. (a) and (b) the object transparency function and its Fourier transform.

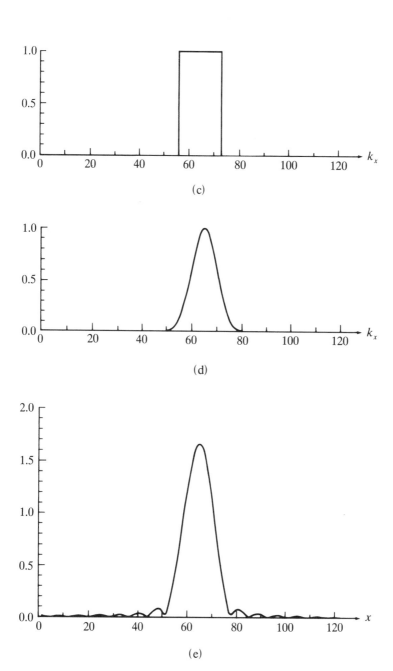

Figure 4.10 (*Continued*) (c) the coherent transfer function. *B* is chosen such that the coherent transfer function spawns the main lobe in (b); (d) and (e) the Fourier transform of the image intensity and the image intensity, respectively, under coherent illumination.

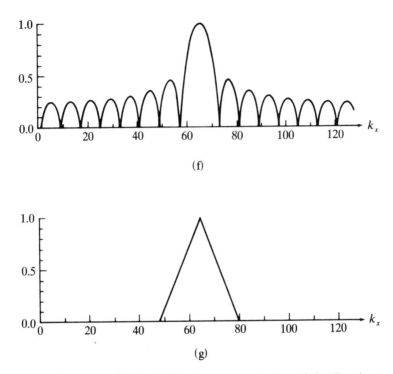

(f)

(g)

Figure 4.10 (*Continued*) (f) and (g) the autocorrelation of the Fourier transform of the object and the OTF, respectively.

coherent and incoherent imaging are plotted in Figure 4.10. Note that the general features show more fringes (coherent speckle noise) in the case of coherent imaging.

4.6 Optical Image Processing

The advantage of having the exact Fourier transform of the input transparency on the common focal plane, or Fourier plane, of a two-lens system as shown in Figure 4.7 makes the setup ideal to do Fourier-plane processing. In what follows, we give some examples of coherent image processing using this method. More examples will appear as problems at the end of this chapter, and in the following chapter.

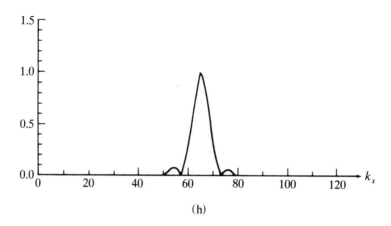

(h)

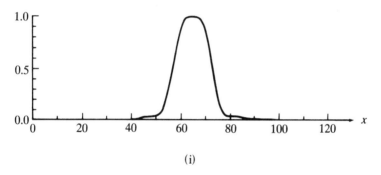

(i)

Figure 4.10 (*Continued*) (h) and (i) the Fourier transform of the image intensity and the image intensity under incoherent illumination. Note that all scales are arbitrary, and only the magnitudes of the transforms are plotted.

Example 4.4 Removing Horizontal Lines from a 2D Grating

The input transparency is a grating represented as

$$t(x, y) = \left(\text{rect}\left(\frac{x}{X}\right) * \sum_{m=-\infty}^{\infty} \delta(x - mX_0) \right)$$

$$\times \left(\text{rect}\left(\frac{y}{Y}\right) * \sum_{n=-\infty}^{\infty} \delta(y - nY_0) \right), \qquad (4.6\text{-}1)$$

as shown in Figure 4.11(a). The field on the common focal plane of the image processing system in Figure 4.7 (assume $f_1 = f_2 = f$ for simplicity) is proportional to the Fourier transform of the input transparency $t(x, y)$,

$$\psi_p(x, y) \propto 4\pi^2 \frac{XY}{X_0 Y_0} \text{sinc}\left(\frac{k_x X}{2\pi}\right) \text{sinc}\left(\frac{k_y Y}{2\pi}\right) \sum_{m=-\infty}^{\infty} \delta\left(k_x - \frac{2\pi m}{X_0}\right)$$

$$\times \sum_{n=-\infty}^{\infty} \delta\left(k_y - \frac{2\pi n}{Y_0}\right)\Bigg|_{\substack{k_x = k_0 x/f \\ k_y = k_0 y/f}}$$

$$= 4\pi^2 \frac{XYf^2}{X_0 Y_0 k_0^2} \text{sinc}\left(\frac{k_0 Xx}{2\pi f}\right) \text{sinc}\left(\frac{k_0 Yy}{2\pi f}\right)$$

$$\times \sum_{m=-\infty}^{\infty} \delta\left(x - \frac{2\pi m f}{k_0 X_0}\right) \sum_{n=-\infty}^{\infty} \delta\left(y - \frac{2\pi n f}{k_0 Y_0}\right). \qquad (4.6\text{-}2)$$

The Fourier-plane field distribution is shown in Figure 4.11(b).

Suppose, now, we desire to see only the vertical lines on the image plane. As we show below, this can be achieved by placing a spatial filter on the Fourier plane of the type

$$s_f(x, y) = \text{rect}\left(\frac{Y}{Y_f}\right), \qquad Y_f < \frac{4\pi f}{k_0 Y_0}. \qquad (4.6\text{-}3)$$

This spatial filter is essentially a rectangular slit in one dimension parallel to the x_f axis and is drawn in Figure 4.11(c). The field to the right of this spatial filter is thus proportional to

$$4\pi^2 \frac{XYf^2}{X_0 Y_0 k_0^2} \text{sinc}\left(\frac{k_0 Xx}{2\pi f}\right) \sum_{m=-\infty}^{\infty} \delta\left(x - \frac{2\pi m f}{k_0 X_0}\right) \delta(y);$$

hence, on the back focal plane of the second lens,

$$\psi_{pi}(x, y) \propto \text{rect}\left(\frac{x}{X}\right) * \sum_{m=-\infty}^{\infty} \delta(x - m X_0), \qquad (4.6\text{-}4)$$

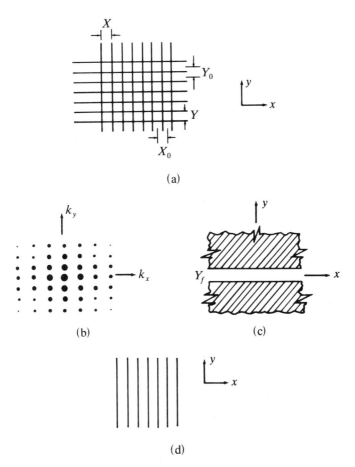

Figure 4.11 Image processing of the transparency function in Eq. (4.6-1).

that is, we retrieve only the vertical lines, as shown in Figure 4.11(d). Horizontal lines may be retrieved similarly, by reorienting the slit along the y-axis.

Before ending this example, we would like to point out that the spatial filter used here had the amplitude characteristics of a one-dimensional low-pass filter. However, we mentioned nothing about the phase characteristics. A discussion on the realization of *complex spatial filters* having both amplitude and phase characteristics is presented in the next chapter.

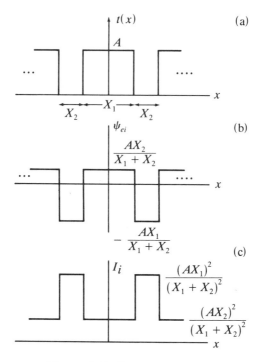

Figure 4.12 Contrast reversal.

Example 4.5 Contrast Reversal

Consider a periodic amplitude transmittance in one dimension, for simplicity, as shown in Figure 4.12(a). Suppose this is placed in the object plane of our two-lens system of Figure 4.7 and a spatial filter in the shape of a tiny opaque stop is introduced on the axis as a processing element. Avoiding all the complicated mathematics this time, we can argue that the spatial filter essentially removes the *dc component* $AX_1/(X_1 + X_2)$ from the object field. Once again, assuming $f_1 = f_2 = f$, the field on the image plane will have the shape shown in Figure 4.12(b). The intensity distribution is plotted in Figure 4.12(c). For $X_1 > X_2$, it is easy to see that the image intensity shows a *contrast reversal* over the object intensity: regions that were previously brighter now appear darker after the image processing. Of course, there is a certain amount of background illumination, and the method fails if $X_1 < X_2$.

✦

Example 4.6 Seeing a Phase Object

It is easy to understand why it is not generally possible to see a phase object of the type

$$t(x) = e^{j\phi(x)}. \qquad (4.6\text{-}5a)$$

This is because the *intensity transmittance*, proportional to $|t(x)|^2$, is equal to a constant. One way of seeing the structure of phase objects is to employ the two-lens system of Figure 4.7 with a small stop on the axis on the Fourier plane as in Example 4.2. Assuming $\phi(x) \ll 1$,

$$t(x) \simeq 1 + j\phi(x), \qquad (4.6\text{-}5b)$$

so that, on the Fourier plane, the field immediately in front of the spatial filter (the small stop) is

$$\psi_p(x, y) \propto \left\{ \frac{2\pi f}{k_0} \delta(x) + j\Phi\left(\frac{k_0 x}{f}\right) \right\} \delta(y), \qquad (4.6\text{-}6)$$

where $\Phi(k_x, k_y) = \mathcal{F}_{xy}[\phi(x, y)]$. The opaque stop now removes the $\delta(x)$, so that

$$\psi_p \propto \Phi\left(\frac{k_0 x}{f}\right) \delta(y). \qquad (4.6\text{-}7)$$

Hence, on the image plane the intensity becomes

$$I_i \propto \phi^2(-x), \qquad (4.6\text{-}8)$$

and the intensity variation shows the nature of the phase structure.

It is also possible to monitor different derivatives of the phase, if desired. For instance, by inserting a Fourier-plane filter with the characteristic

$$s_f = ax^2, \qquad (4.6\text{-}9)$$

it is easy to show that the intensity variation on the image plane is

$$I_i \propto \left| \frac{d^2 \phi(-x)}{dx^2} \right|^2. \qquad (4.6\text{-}10)$$

For other illustrations of coherent image processing, the reader is referred to the problems that follow and to the next chapter.

Problems

4.1   Starting from Eq. (4.2-10) with $p_f(x, y) = 1$, show that ψ_p at $z = d_i$ reduces to a form similar to Eq. (4.2-6).

4.2   Assume that a Gaussian beam with a waist radius w_0,

$$\psi_e(x, y) = A \exp\left(-\frac{x^2 + y^2}{w_0^2} \right),$$

travels a distance d to a lens with a focal length f. Calculate the radius of the beam at the back focal plane of the lens.

4.3   Often, the beam emerging from a laser diode has an elliptical Gaussian cross section,

$$\psi_e(x, y) = A \exp\left[-\left(\frac{x^2}{w_{0x}^2} + \frac{y^2}{w_{0y}^2} \right) \right],$$

instead of a circular Gaussian cross section. If a lens of focal length f (> 0) is placed right next to the exit aperture of the laser diode, calculate the distance z_0 where the on-axis amplitude of the beam will be a maximum.

4.4   An input transparency $t(x, y)$ is placed at a distance $d < f$ behind a lens of focal length f. Find the field distribution at the back focal plane if a plane wave of unit amplitude is incident on the lens.

4.5   A monochromatic plane wave is obliquely incident on a transparency $t(x, y)$ placed on the front focal plane of the converging lens L_1 of the optical system shown in Figure 4.7. Assume $f_1 = f_2 = f$.

(a) Calculate the complex field at the back focal plane of L_1.

(b) If a small pinhole is placed on this placed on the lens axis, predict the shape of the intensity pattern on the back focal plane of the second lens L_2 as θ varies.

4.6   Consider an optical system like that shown in Figure 4.7, with $f_1 = f_2 = f$. An object with the amplitude transmission function

$$f(x) = \cos(2\pi x/a) + \cos(2\pi x/b), \qquad a \neq b,$$

is placed in the front focal plane of L_1. In the back focal plane of the lens L_2, we want to obtain a field pattern proportional to $\cos(2\pi x/a)$. How do we accomplish such a task? Perform the mathematical analysis in one transverse dimension.

4.7   Consider the image processing scheme of Figure 4.7.

(a) What is the intensity distribution $I(x, y)$ on the observation plane in terms of the input transparency function $t(x, y)$, assuming that there is no spatial filter placed on the common plane? Give *simple* reasons for your answer.

(b) With $f_1 = f_2 = f$, suppose that $t(x, y) = \exp[-j\phi(x, y)]$. Write down the expression for $I(x, y)$.

(c) With a transparency function as in (b), suppose now that a spatial filter $s_f(x, y) = a(x^2 + y^2)$ is introduced on the common focal plane between the two lenses. How is the resulting intensity related to the object phase? Do not make any approximations on $t(x, y)$.

4.8   Given a periodic transparency function

$$t(x) = \sum_{n=-\infty}^{\infty} T_n e^{jnKx},$$

where T_n denotes the Fourier coefficients and $K = 2\pi/X$, where X is the repetition period, write down, using the transfer function concept of propagation, the spectrum of the complex field a distance z behind the transparency if it is illuminated with a plane wave.

Hence, find at what distance $z = z_i$ the complex field becomes exactly the same as the input transparency function. (This is called a case of *lensless imaging*!)

4.9 For a two-lens image processing system shown in Figure 4.7, show that the field on the back focal plane of the second lens L_2 for an input transparency $t(x, y)$ at the front focal plane of L_1 is given by Eq. (4.2-15) and the image intensity by Eq. (4.2-16), where $k(x, y)$ is defined in Eq. (4.3-4) and with $M = -f_2/f_1$.

4.10 Derive Eqs. (4.5-8a)–(4.5-8c) by starting from Eq. (4.5-5a) [or (4.5-5(b)] and taking its Fourier transform.

4.11 Derive the two properties of the OTF, as given in Eqs. (4.5-9) and (4.5-10) by starting from the definition of the OTF, given in Eq. (4.5-8c).

4.12 For a circular pupil function $p_f(x, y) \triangleq p_f(r) = \text{circ}(r/r_0)$, find and plot the following for a single-lens imaging system:
(a) the coherent transfer function;
(b) the OTF.

4.13 Referring to the two-lens coherent image-processing system of Figure 4.7 and assuming $f_1 = f_2 = f$, suppose that a transparency $t_1(x, y)$ is placed on the object plane. For the spatial filter $s_f(x, y)$ of the form
(a) $s_f(x, y) = T_2(k_0 x/f, k_0 y/f)$,
(b) $s_f(x, y) = T_2^*(-k_0 x/f, -k_0 y/f)$,
where $T_2(k_x, k_y) = \mathscr{F}_{xy}[t_2(x, y)]$, find the expressions for the image intensity on the image plane in terms of $t_1(x, y)$ and $t_2(x, y)$. What type of mathematical operations does this system represent?

4.14 Consider the single-lens image processing system shown in Figure P4-14. Assuming a spatial filter $s_f(x, y)$ is placed on the back focal plane of the lens, find, from first principles, the response of the imaging system to a shifted δ function input, namely, $\delta(x - x_o, y - y_o)$.

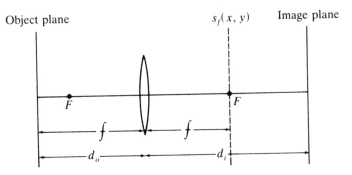

Figure P4.14

References

4.1 Beran, M. J. and G. B. Parrent (1964). *Theory of Partial Coherence*. Prentice-Hall, Englewood Cliffs, N.J.

4.2 Ghatak, A. K. and K. Thyagarajan (1978). *Contemporary Optics*. Plenum, New York.

4.3 Glaser, I. (1987). In *Progress in Optics*, Vol. 24 (E. Wolf, ed.) Chapter V. North-Holland, Amsterdam.

4.4 Goodman, J. W. (1968). *Introduction to Fourier Optics*. McGraw-Hill, New York.

4.5 Lohmann, A. W. and W. T. Rhodes (1978). *Appl. Opt.* **17** 1411.

4.6 Poon, T.-C. and A. Korpel (1979). *Opt. Lett.* **10** 317.

4.7 Stoner, W. (1978). *Appl. Opt.* **16** 265.

4.8 Yu, F. T. S. (1983). *Optical Information Processing*. Wiley, New York.

Chapter 5 — Optical Information Processing (II)

In the previous chapter, we discussed the Fourier-transforming and imaging properties of lenses, the frequency response of imaging systems under coherent and incoherent illumination, and simple image-processing systems. Recall that image processing involves placing a suitable Fourier-plane spatial filter. For instance, in Example 4.5, we employed a spatial *low-pass filter* (at least in amplitude) to achieve the required processing. Those familiar with linear systems will remember that an ideal low-pass filter has not only a specified amplitude characteristic, but a phase characteristic as well. This leads us to the question of how we can realize *complex* spatial filters in general for image-processing purposes. In this chapter, we will address this topic along with a related area called holography. In that connection, we will briefly introduce the properties of photographic films that are usually used in the construction of spatial filters and holograms. We will not include the traditional discussion on photometry and related topics in this book. Interested readers are referred to Nicodemus (1983).

5.1 Characteristics of Photographic Films

Every photographic film comprises a transparent *base* with a coating of a *photosensitive emulsion* of silver halide on top. When the film is exposed to light, the silver halide changes to metallic silver, which is opaque at optical frequencies. The process of *fixing* removes the unexposed halide by dissolving it away; hence, those parts of the film that receive no light become transparent after the fixing process. The entire process is much more gradual than it sounds, and parts of the film receiving more light appear darker than the parts that receive less light. This analog nature comes from the fact that the halides are deposited as micrograins on the transparent base, so that a certain amount of light energy can only enable the chemical reaction to occur in some of these grains.

The transparency produced after fixing is called the *negative*, for reasons that are obvious from the preceding paragraph. When the negative is exposed to light, more light is transmitted through those regions on the negative that are more transparent. A figure of merit for this is the intensity transmittance

$$\tau_n(x, y) \triangleq \text{average}\left\{ \frac{\text{intensity transmitted at } (x, y)}{\text{intensity incident at } (x, y)} \right\}, \quad (5.1\text{-}1)$$

where the subscript n signifies the fact that we have a *negative* transparency. The average is taken over several film micrograins.

During the exposure process, the greater the amount of light incident on the photographic film, the more silver is formed, and hence, τ_n is comparatively lower. That is, τ_n decreases with increasing *exposure* $\mathscr{H}$ defined as the product of the exposure time T and incident light intensity $I(x, y)$. A standard procedure for quantifying all this is to define the *photographic density* $\mathscr{D}$,

$$\mathscr{D}(x, y) \triangleq \log_{10}(1/\tau_n). \quad (5.1\text{-}2)$$

The variation of $\mathscr{D}$ with $\log_{10} \mathscr{H}$ is usually specified for every film. Typical variations, called the *Hurter–Driffield* (H & D) *curves*, are shown in Figure 5.1. Note that the solid curve has a linear region with a slope γ_n, which is most useful for recording purposes. The γ_n depends on the following:

(a) *the type of film emulsion (i.e., the silver halide);*
(b) *the type of developer used;*
(c) *the developing time.*

Optical Information Processing (II)

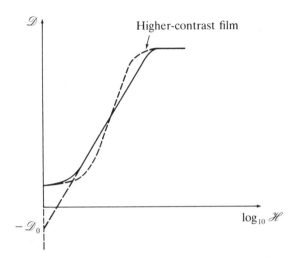

Figure 5.1 The H & D curves of typical films.

Based on the type of film emulsion, we can have negatives of differing *contrasts*. A higher-contrast film has a larger value of γ_n, as shown by the dashed curve in Figure 5.1.

In the linear region (see Figure 5.1),

$$\mathscr{D} = \gamma_n \log_{10} \mathscr{H} - \mathscr{D}_0 = \gamma_n \log_{10} IT - \mathscr{D}_0, \qquad (5.1\text{-}3)$$

so that, from Eqs. (5.1-2) and (5.1-3),

$$\tau_n(x, y) = K_n[I(x, y)]^{-\gamma_n}, \qquad K_n = 10^{+\mathscr{D}_0} T^{-\gamma_n}, \qquad (5.1\text{-}4)$$

where $I(x, y)$ is the (local) intensity of light to which the photographic film was exposed.

Often, it is desirable to obtain a *positive* transparency, which is a replica of the object. To achieve this, we photograph the first negative transparency (say, τ_{n1}), producing a second negative transparency. This resulting transparency is now a positive copy of the original object and is characterized by an *intensity* transmittance

τ_p, where

$$\tau_p(x, y) = K_{n2}\left(K_{n1}[I(x, y)]^{-\gamma_{n1}}\right)^{-\gamma_{n2}}$$

$$\triangleq K_p[I(x, y)]^{\gamma_p}, \qquad \gamma_p = \gamma_{n1}\gamma_{n2}, \qquad (5.1\text{-}5)$$

where K_p is a constant and γ_p represents the overall gamma (γ) for the two-step process. In Eq. (5.1-5), the subscripts 1 and 2 refer to the first and second stages involved in producing the final transparency.

   If this positive transparency is to be used for image processing in a coherent system (which is our interest in this chapter), we have to know its *amplitude* transmittance $t_p(x, y)$. This is given by

$$t_p(x, y) = \left[\tau_p(x, y)\right]^{1/2} e^{-j\phi(x, y)}, \qquad (5.1\text{-}6)$$

where we have included the extra phase term $\phi(x, y)$ to take into account possibly varying thickness of the final transparency at different parts. Without getting into too much detail, we remark that the phase term can be effectively removed by immersing the final transparency in a so-called index-matching fluid confined between two parallel optical flats, as shown in Figure 5.2. In the rest of this chapter, we will therefore neglect the effect(s) of $\phi(x, y)$.

   We mention at this point that, for most image-processing or holographic applications, an overall γ ($= \gamma_p$) of 2 is desired. This means that the amplitude transmittance $t_p(x, y)$ of the positive transparency is, according to Eqs. (5.1-5) and (5.1-6),

$$t_p(x, y) \propto I(\acute{x}, y), \qquad (5.1\text{-}7)$$

where $I(x, y)$ is proportional to the intensity of the light incident on the original film from which τ_{n1} was constructed.

Optical Information Processing (II)

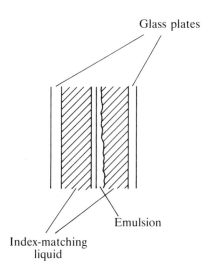

Glass plates

Emulsion

Index-matching
liquid

Figure 5.2 A technique to remove phase distortion from a positive transparency.

5.2 Notion of a Reference Wave to Record Amplitude and Phase; Holography

As mentioned in the preamble to this chapter, the construction of complex spatial filters requires the recording of both amplitude and phase information. Although recording the amplitude
information is fairly straightforward, recording the phase information is more involved because, by Eq. (5.1-7), it is only possible to
make the amplitude transmittance proportional to the intensity
information. However, although it is not possible to record the
absolute phase, *relative* phases can be recorded. By relative phase,
we mean the *phase difference* between a certain wave and a *reference wave*. The notion of the reference wave is extremely important
in connection with the construction of complex spatial filters and
holograms, as will be seen later.

In order to see how the reference wave can give relative
phase information, consider Young's double-slit experiment (Example 3.11) in a slightly different light. In that example, we were
interested in the Fraunhofer diffraction pattern of two slits. Suppose, instead, we now have two point sources and wish to determine the complex field at an arbitrary distance z. We can formally
use the Fresnel diffraction formula, or alternatively, realize that

these point sources A and B (see Figure 3.18) emanate spherical waves, so that the field on the observation plane is (after Example 3.4),

$$\psi_p = \frac{jk_0 \exp(-jk_0 z)}{2\pi z} \exp\left[-j\frac{k_0}{2z}(x^2 + y^2)\right]$$

$$\times \left[\psi_{0A} \exp\left[j\left(\frac{k_0 d}{2z}x - \phi\right)\right] + \psi_{0B} \exp\left(-j\frac{k_0 d}{2z}x\right)\right]$$

$$\times \exp\left(\frac{-jk_0}{8z}d^2\right). \tag{5.2-1}$$

In writing Eq. (5.2-1), we have assumed the relative strengths of the point sources to be ψ_{0A} and ψ_{0B}, and the phase of the sources at A and B to be $-\phi$ and 0, respectively.

If we now place a photographic plate on the observation plane and construct a positive transparency with an overall γ of 2, we can express the amplitude transmittance of this transparency, using Eqs. (5.1-7) and (5.2-1), as

$$t_p(x, y) \propto \frac{k_0^2}{4\pi^2 z^2}\left\{\psi_{0A}^2 + \psi_{0B}^2 + \psi_{0A}\psi_{0B} \exp\left[+j\left(\frac{k_0 d}{z}x - \phi\right)\right]\right.$$

$$\left. + \psi_{0A}\psi_{0B} \exp\left[-j\left(\frac{k_0 d}{z}x - \phi\right)\right]\right\}$$

$$= \frac{k_0^2}{4\pi^2 z^2}\left\{\psi_{0A}^2 + \psi_{0B}^2 + 2\psi_{0A}\psi_{0B} \cos\left(\frac{k_0 d}{z}x - \phi\right)\right\}. \tag{5.2-2}$$

Note that the phase information of the source at A has been recorded along with its amplitude information ψ_{0A} in the positive transparency. This was made possible by the presence of the second source at B, which we, from now on, will call the *reference*. Thus, from this very elementary example, we conclude that a *reference wave* is important if we want to record the amplitude and phase of a given complex field. This is essential in the construction of complex spatial filters. Examples of such filters and their applications will follow in Section 5.4.

Let us now examine the effect of placing the positive transparency $t_p(x, y)$, as just constructed, in the position of the photographic plate (observation plane) in Figure 3.18 and illuminating it with a point source of strength ψ_{0C} located at B (where the reference source was placed). The optical field incident on the transparency from the left is

$$\psi_{pL} = \frac{jk_0\psi_{0C}}{2\pi z} \exp(-jk_0 z)\exp\left\{-j\frac{k_0}{2z}\left[\left(x + \frac{d}{2}\right)^2 + y^2\right]\right\}. \quad (5.2\text{-}3)$$

The field ψ_{pR} immediately behind the transparency is obtained by multiplying Eq. (5.2-3) with Eq. (5.2-2). We shall treat a simple case with $\phi = 0$ to bring out the essential point. Instead of writing the entire expression for ψ_{pR}, we consider the term of interest to us for this discussion, namely,

$$\psi_{pR \, rel} = \frac{jk_0^2\psi_{0A}\psi_{0B}\psi_{0C}}{8\pi^3 z^3} \exp(-jk_0 z)$$

$$\times \exp\left\{-j\frac{k_0}{2z}\left[\left(x + \frac{d}{2}\right)^2 + y^2\right]\right\}\exp\left(j\frac{k_0 d}{z}x\right), \quad (5.2\text{-}4a)$$

where the first two exponentials come from Eq. (5.2-3) and the third exponential comes from Eq. (5.2-2). Equation (5.2-4a) can be reexpressed as

$$\psi_{pR \, rel} = \frac{\psi_{0B}\psi_{0C}k_0}{4\pi^2 z^2}\left[\frac{jk_0\psi_{0A}}{2\pi z} \exp(-jk_0 z)\right.$$

$$\left.\times \exp\left\{-j\frac{k_0}{2z}\left[\left(x - \frac{d}{2}\right)^2 + y^2\right]\right\}\right]. \quad (5.2\text{-}4b)$$

Note that Eq. (5.2-4b) represents a spherical wave, appearing to come from the point A, which was our original *object point* (see Figure 5.3). In fact, by looking into the transparency from the right, we see a *virtual image* of the object, which appears at exactly the same location where the original object was placed. The transparency t_p [as in Eq. (5.2-2)] is called the *hologram* of the point-

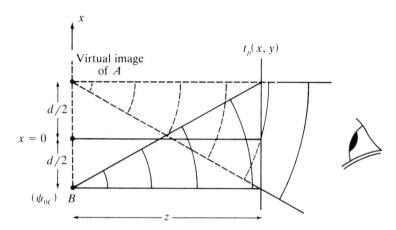

Figure 5.3 Reconstruction from the hologram of a point object using a point-source reading field.

object A. The word "hologram" refers to the recording of the *whole* (*holos* in Greek) information pertaining to the object, that is, the amplitude as well as the phase. We can see the object by looking into the hologram when it is illuminated by the *reading* or *reconstruction* wave of strength ψ_{0C}. Note that the intensity of the *observed image* is proportional to ψ_{0B} and ψ_{0C}, the strengths of the reference and reading waves, respectively; hence, it helps to have strong reference and reading waves.

As a second example of recording and reconstruction, consider the recording of the hologram of a point source, using a plane wave as the reference, as shown in Figure 5.4(a). Assuming the relative phase difference between the point source and the plane wave to be $-\phi$ as before, it is easy to check that the complex field phasor on the recording medium is

$$\psi_p = \exp(-jk_0 z)\left[\frac{jk_0}{2\pi z}\psi_{0A}\exp\left[-j\left\{\frac{k_0}{2z}(x^2 + y^2) + \phi\right\}\right] + \psi_{0B}\right],$$

$$(5.2\text{-}5)$$

where ψ_{0A} is the strength of the point source at A and ψ_{0B} is the amplitude of the plane wave. The positive transparency, constructed from the film placed on the observation plane, with an

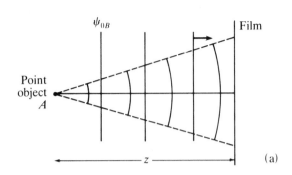

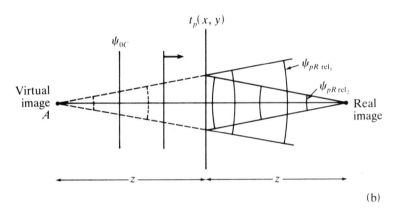

Figure 5.4 (a) Recording and (b) reconstruction of a point object using a
plane wave.

overall γ of 2, has the amplitude transmittance

$$t_p(x, y) \propto \frac{k_0^2}{4\pi^2 z^2}\psi_{0A}^2 + \psi_{0B}^2 + \frac{k_0}{\pi z}\psi_{0A}\psi_{0B}$$

$$\times \sin\left\{\frac{k_0}{2z}(x^2 + y^2) + \phi\right\}, \qquad (5.2\text{-}6)$$

showing once again that both the amplitude and the phase information, so vital for the realization of a complex filter, are recorded.

To see how the reconstruction works in this case, assume, as before, that the hologram is placed in the position of the photographic plate and illuminated with a plane wave of amplitude ψ_{0C} whose phase is zero at the plane where the original object was located. Then the complex field illuminating the transparency is $\psi_{0C}e^{-jk_0z}$, which can be multiplied with Eq. (5.2-6) to obtain the field immediately behind the transparency. With $\phi = 0$, we can isolate the following relevant terms:

$$\psi_{pR\,rel_1} = \psi_{0B}\psi_{0C}\left[\frac{jk_0}{2\pi z}\psi_{0A}\exp(-jk_0z)\exp\left(\frac{-jk_0}{2z}(x^2+y^2)\right)\right];$$

$$(5.2\text{-}7a)$$

$$\psi_{pR\,rel_2} = \psi_{0B}\psi_{0C}\left[\frac{-jk_0}{2\pi z}\psi_{0A}\exp(-jk_0z)\exp\left(\frac{jk_0}{2z}(x^2+y^2)\right)\right].$$

$$(5.2\text{-}7b)$$

The first term, $\psi_{pR\,rel_1}$, is similar to Eq. (5.2-4b) and represents a spherical wave that appears to diverge from point A and, hence, contributes to a virtual image of the object. On the other hand, Eq. (5.2-7b) is the expression of a converging wave front that comes to a focus at a distance z behind the hologram and is responsible for a real image of the object. Thus, if the recording and reconstruction are done using a plane wave, we can get both the real and virtual images. This is shown in Figure 5.4(b). The practical disadvantage of this scheme is that the wave contributions $\psi_{pR\,rel_1}$ and $\psi_{pR\,rel_2}$ overlap with each other. This problem is easily resolved using the setup shown in the following section.

We end this discussion with the remark that a reexamination of Eq. (5.2-7) reveals that whereas the term proportional to $\exp[-j(k_0/2z)(x^2+y^2)]$ is multiplied by ψ_{0A}, the term $\exp[j(k_0/2z)(x^2+y^2)]$ is multiplied by ψ_{0A}^* in the event that the object is complex ($\phi \neq 0$). Also note that the preceding procedure for recording and reconstruction may be extended to an ensemble of point sources in three dimensions, acting as the object.

5.3 Construction of Practical Holograms or Complex Filters

In this section, we are going to consider two examples of the construction of holograms or complex filters. One of the first holograms that was constructed is the *Gabor hologram* in which the object transparency, from which the hologram is constructed, is assumed to have a strong quiescent transmittance. This means that if $t'(x, y)$ is really the information of which the hologram is to be constructed, the transmittance of the transparency $t(x, y)$ that is illuminated by a coherent source should be

$$t(x, y) = t_0 + t'(x, y), \qquad (5.3\text{-}1)$$

where $t_0 \gg |t'(x, y)|$. In other words, $t(x, y)$ is assumed to be highly transmissive. This arrangement is necessary in order to provide a strong reference beam. A more practical setup is the Leith–Upatneiks technique, which is shown in Figure 5.5. The lens L forms a coherent beam from the point source, part of which illuminates the object $t(x, y)$. A prism bends the other part of the coherent beam through an angle θ and helps provide the reference, which is assumed, for simplicity, to remain virtually undiffracted during its passage to the plane of the film. The two fields, one from the object and assumed equal to $\psi_{0A}(x, y)$ at the plane of the film [which is the Fresnel diffraction of $t(x, y)$] and the other forming the reference and equal to $\psi_{0B} \exp[jk_0(\sin \theta)y]$ (ψ_{0B} real), inter-

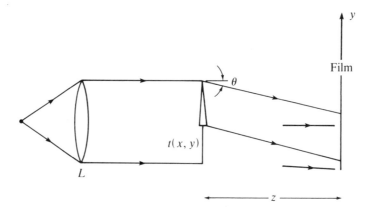

Figure 5.5 Schematic of the recording of a Leith–Upatneiks hologram.

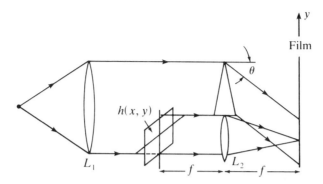

Figure 5.6 Construction of a complex spatial filter.

fere on the plane of the film. Any relative phase difference between these two fields, due to varying propagation distances, can be included in a complex constant K multiplying ψ_{0A}. We also assume that the amplitudes of the optical field incident on the object $t(x, y)$ and the prism are equal. If a positive transparency $t_p(x, y)$ with an overall γ of 2 is once again constructed, then

$$t_p(x, y) \propto \psi_{0B}^2 + |K|^2 |\psi_{0A}(x, y)|^2$$

$$+ K\psi_{0B}\psi_{0A}(x, y)\exp[-jk_0(\sin \theta)y]$$

$$+ K^*\psi_{0B}\psi_{0A}^*(x, y)\exp[+jk_0(\sin \theta)y]. \quad (5.3\text{-}2)$$

Equation (5.3-2) therefore gives the amplitude transmittance of the *Leith–Upatneiks hologram* constructed from the object $t(x, y)$.

As a variation of the setup of the Leith–Upatneiks hologram, consider the arrangement shown in Figure 5.6, with which we now intend to record the Fourier transform of the object. This is achieved by the lens L_2, which projects the Fourier transform of a transparency $h(x, y)$ on the plane of the film. The hologram will now have a transparency function

$$t_p(x, y) \propto \psi_{0B}^2 + |K_1|^2 |H|^2 + K_1\psi_{0B}H \exp[-jk_0(\sin \theta)y]$$

$$+ K_1^*\psi_{0B}H^* \exp[+jk_0(\sin \theta)y], \quad (5.3\text{-}3a)$$

where

$$H = H\left(\frac{k_0 x}{f}, \frac{k_0 y}{f}\right) = \mathscr{F}_{xy}\{h(x, y)\}\Big|_{\substack{k_x = k_0 x/f \\ k_y = k_0 y/f}} \quad (5.3\text{-}3b)$$

and where K_1 is once again a complex constant. Thus, the hologram or *complex filter* contains the magnitude and phase information of the Fourier transform of the transparency $h(x, y)$.

5.4 Reconstruction and Applications of Holograms and Complex Filters

In this section, we investigate the effects of illuminating the hologram and the complex filter just discussed with different complex optical fields. First, consider the Leith–Upatneiks hologram as in Eq. (5.3-2) to be illuminated by a collimated beam, as shown in Figure 5.7. Relevant terms in the expression for the optical field behind the hologram will be proportional to the third and fourth terms in Eq. (5.3-2). Recall that ψ_{0A} is the complex field, due to diffraction of the light after it passes through the original transparency $t(x, y)$, on the plane of the film. Hence, using Eqs. (3.3-19a), (3.3-19b) and (5.3-2), the first relevant term, that is, the third term

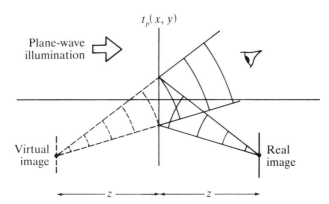

Figure 5.7 Reconstruction from the Leith–Upatneiks hologram using a plane wave.

in Eq. (5.3-2), is, explicitly,

$$\psi_{pR\,rel_1} = \frac{jk_0 K}{2\pi z}\psi_{0B}\exp[-jk_0(z + y\sin\theta)]$$

$$\times \int\int_{-\infty}^{\infty} t(x', y')\exp\left\{\frac{-jk_0}{2z}\left[(x - x')^2 + (y - y')^2\right]\right\} dx'\,dy'.$$

$$(5.4\text{-}1)$$

In the context of reconstruction, this term [which is similar to Eq. (5.2-7a)] would, therefore, give rise to a virtual image, as shown in Figure 5.7, because diverging spherical waves appear to emanate from the location of the virtual image, with the nominal direction of propagation at an angle θ to the axis. Correspondingly, the fourth term in Eq. (5.3-2) is

$$\psi_{pR\,rel_2} = -\frac{jk_0 K^*}{2\pi z}\psi_{0B}\exp[-jk_0(z - y\sin\theta)]$$

$$\times \int\int_{-\infty}^{\infty} t^*(x', y')\exp\left\{\frac{+jk_0}{2z}\left[(x - x')^2 + (y - y')^2\right]\right\} dx'\,dy'.$$

$$(5.4\text{-}2)$$

Note that this expression represents a collection of converging spherical waves with the nominal direction of propagation at an angle $-\theta$ to the axis. We now show that this complex field gives rise to a real image as shown in Figure 5.7, by employing the transfer function concept of propagation, as enunciated in Eq. (3.3-16). The Fourier transform of Eq. (5.4-2) is proportional to $T^*(-k_x, -k_y)\exp[-j(k_x^2 + k_y^2)z/2k_0]$, where we have disregarded the exponential before the integral for simplicity; hence, after propagation through a distance z', the Fourier transform of the complex field must be $T^*(-k_x, -k_y)\exp\{-j[(k_x^2 + k_y^2)/2k_0](z - z')\}$. When $z' = z$, the complex field, therefore, will be $t^*(x, y)$, and hence the intensity will be proportional to $|t(x, y)|^2$. We end this discussion by stating that if a screen is placed perpendicular to

Optical Information Processing (II)

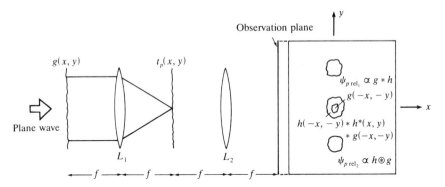

Figure 5.8 Application of a complex filter constructed as shown in Figure 5.6. The exact mathematical expressions for $\psi_{p\,\text{rel}_1}$ and $\psi_{p\,\text{rel}_2}$ appear in Eqs. (5.4-4) and (5.4-5).

the nominal direction of propagation of $\psi_{pR\,\text{rel}_1}$, we will observe the Fresnel diffraction pattern of the original transparency $t(x, y)$.

Next, let us look at an application of the complex filter or hologram described by Eqs. (5.3-3a) and (5.3-3b). Assume, for instance, that the complex filter is inserted on the common focal plane of two lenses L_1 and L_2 as shown in Figure 5.8. For convenience, assume the focal lengths of these lenses to be equal to the focal length f of the lens that was used in constructing the complex filter. Now let a transparency $g(x, y)$ be placed in the front focal plane of L_1 and illuminated with a plane wave. Then the field incident on the frequency plane mask is $K_2 G(k_0 x/f, k_0 y/f)$, where K_2 is a complex constant. This gets multiplied with the transparency function $t_p(x, y)$ of the complex filter as in Eq. (5.3-3a). Hence, on the back focal plane of L_2, the complex field is

$$\psi_p(x, y) \propto \left[\mathscr{F}_{xy}\left\{ K_2 G\left[\psi_{0B}^2 + |K_1|^2 |H|^2 \right. \right. \right.$$

$$+ K_1 \psi_{0B} H \exp(-jk_0(\sin\theta)y)$$

$$\left. \left. \left. + K_1^* \psi_{0B} H^* \exp(jk_0(\sin\theta)y) \right] \right\} \right] \Big|_{\substack{k_x = k_0 x/f \\ k_y = k_0 y/f}}. \quad (5.4\text{-}3)$$

Once again, the relevant terms are the third and fourth terms in

Eq. (5.4-3). For instance, we can express the third term as

$$\psi_{p\,\mathrm{rel}_1} \propto \mathscr{F}_{xy}\left[G\left(\frac{k_0 x}{f}, \frac{k_0 y}{f}\right) H\left(\frac{k_0 x}{f}, \frac{k_0 y}{f}\right)\right.$$

$$\left.\left.\times \exp\left(-jk_0(\sin\theta)y\right)\right]\right|_{\substack{k_x = k_0 x/f \\ k_y = k_0 y/f}}$$

$$\propto g(-x, -y) * h(-x, -y) * \delta(x, y - f\sin\theta)$$

$$= g(-x, -y) * h(-x, -(y - f\sin\theta)), \qquad (5.4\text{-}4)$$

whereas the fourth term yields

$$\psi_{p\,\mathrm{rel}_2} \propto h(-x, -y) \circledast g(-x, -(y + f\sin\theta)), \qquad (5.4\text{-}5)$$

according to our definition of correlation in Eq. (1.2-12). Notice that these two fields are centered around $y = \pm f\sin\theta$, respectively, on the observation plane. Also, note that the first and second terms in Eq. (5.4-3) contribute to complex fields of the forms $g(-x, -y)$ and $h(-x, -y) * h^*(x, y) * g(-x, -y)$, both of which are centered around $y = 0$. This is shown in Figure 5.8. For spatial separation of the *undeviated* light and the convolution and correlation, it is imperative to make θ sufficiently large.

We can use the scheme just presented to implement a *matched filter*. For instance, note from communication theory that if $g(x, y)$ is the input to a matched filter having a transfer function $H(k_x, k_y) = G^*(k_x, k_y)$, where $G(k_x, k_y)$ is the Fourier transform of $g(x, y)$, then the output of the matched filter is $GG^*(k_x, k_y)$, which translates to the autocorrelation of $g(x, y)$ in the spatial domain. In optics, the realizable equivalent of the complex filter $H(k_x, k_y)$ is $t_p(x, y)$, as specified in Eqs. (5.3-3a) and (5.3-3b), and which, as just shown, is adequate to perform the correlation desired.

The matched filter just discussed is one of the major components of a *character-recognition system*, which, in principle, may be described as follows. An input set of characters,

$$g(x, y) = \sum_{i=1}^{N} g_i(x, y), \qquad (5.4\text{-}6)$$

applied to the character-recognition machine with a matched filter having the transfer function

$$H(k_x, k_y) = G_M^*(k_x, k_y), \qquad (5.4\text{-}7a)$$

where

$$G_M(k_x, k_y) = \mathscr{F}_{xy}[g_M(x, y)], \qquad 1 \leq M \leq N, \quad (5.4\text{-}7b)$$

produces the output

$$s(x, y) = \left\{ \sum_{\substack{i=1 \\ i \neq M}}^{N} g_i(x, y) \circledast g_M(x, y) \right\} + g_M(x, y) \circledast g_M(x, y)$$

$$\triangleq \sum_{\substack{i=1 \\ i \neq M}}^{N} R_{iM} + R_{MM}, \qquad (5.4\text{-}8)$$

comprising the autocorrelation of g_M and several cross-correlations. Now,

$$|R_{iM}(0, 0)|^2 = \left| \int\int_{-\infty}^{\infty} g_i(x, y) g_M^*(x, y) \, dx \, dy \right|^2$$

$$\leq \int\int_{-\infty}^{\infty} |g_i(x, y)|^2 \, dx \, dy \int\int_{-\infty}^{\infty} |g_M(x, y)|^2 \, dx \, dy, \quad (5.4\text{-}9)$$

by employing the *Schwartz inequality*. If the total "power" content of all g_is are approximately equal, that is,

$$\int\int_{-\infty}^{\infty} |g_i(x, y)|^2 \, dx \, dy \approx \int\int_{-\infty}^{\infty} |g_j(x, y)|^2 \, dx \, dy, \qquad (5.4\text{-}10)$$

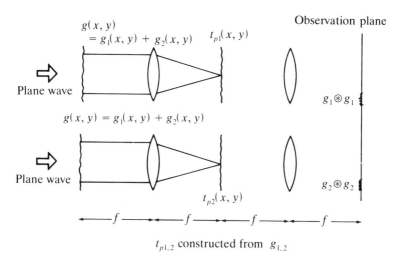

Figure 5.9 A character-recognition scheme.

then Eq. (5.4-9), with the definitions of R_{iM} and R_{MM} as in Eq. (5.4-8), can be readily simplified to

$$|R_{iM}(0,0)| \leq R_{MM}(0,0). \tag{5.4-11}$$

This implies that the amplitude each of the cross-correlation in Eq. (5.4-8) will be less than that of the autocorrelation, facilitating easy detection of the autocorrelation and, hence, the required "character" in the character-recognition problem. A simple character-recognition system that can be used to detect two characters (say g_1, g_2) in a transparency $g(x, y)$ as in Eq. (5.4-6) is illustrated in Figure 5.9.

5.5 Holographic Magnification

Wavefront reconstruction is, in general, three-dimensional in nature. In what follows, we study the lateral and longitudinal magnifications of the holographic image. It is sufficient to use point sources for reference and reconstruction and to restrict ourselves to point objects, because a three-dimensional object can be represented as an ensemble of points.

Consider the geometry for recording shown in Figure 5.10(a). The two point objects 1 and 2 and the reference source 3 emit spherical waves that, on the plane of the film, contribute to complex

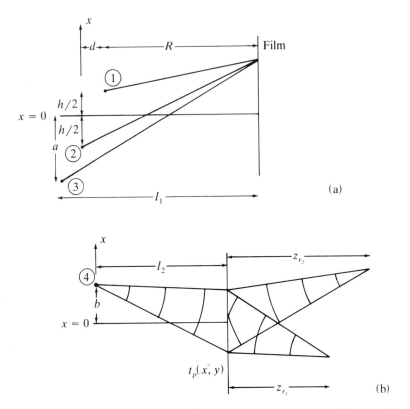

Figure 5.10 Geometry to find lateral and longitudinal holographic magnifications: (a) recording; (b) reconstruction.

fields ψ_{p1}, ψ_{p2}, and ψ_{p3}, respectively, given by

$$\psi_{p1} = A_1 \exp\left(-jk_1\left\{R + \frac{1}{2R}\left[\left(x - \frac{h}{2}\right)^2 + y^2\right]\right\}\right), \qquad (5.5\text{-}1a)$$

$$\psi_{p2} = A_2 \exp\left(-jk_1\left\{(R + d) + \frac{1}{2(R + d)}\left[\left(x + \frac{h}{2}\right)^2 + y^2\right]\right\}\right),$$

$$(5.5\text{-}1b)$$

$$\psi_{p3} = A_3 \exp\left(-jk_1\left\{l_1 + \frac{1}{2l_1}\left[(x + a)^2 + y^2\right]\right\}\right), \qquad (5.5\text{-}1c)$$

where A_1, A_2, and A_3 are complex constants, and where we have assumed k_1 ($= 2\pi/\lambda_1$) to represent the wave number of the three waves. These three waves interfere on the plane of the film, which if recorded and converted into a positive transparency with an overall γ of 2, yields the hologram

$$t_p(x, y) \propto (\psi_{p1} + \psi_{p2} + \psi_{p3})(\psi_{p1}^* + \psi_{p2}^* + \psi_{p3}^*). \quad (5.5\text{-}2)$$

Rather than write down the expression of $t_p(x, y)$ explicitly, we will, on the basis of our previous experience, pick out some relevant terms. The terms of interest are

$$t_{p\,\mathrm{rel}_1} = \psi_{p1}^* \psi_{p3}$$

$$= A_1^* A_3 \exp\left(+jk_1 \left\{ (R - l_1) + \frac{1}{2R}\left[\left(x - \frac{h}{2} \right)^2 + y^2 \right] \right. \right.$$

$$\left. \left. - \frac{1}{2l_1}\left[(x + a)^2 + y^2 \right] \right\} \right), \quad (5.5\text{-}3\mathrm{a})$$

$$t_{p\,\mathrm{rel}_2} = \psi_{p2}^* \psi_{p3}$$

$$= A_2^* A_3 \exp\left(+jk_1 \left\{ (R + d - l_1) + \frac{1}{2(R + d)} \right. \right.$$

$$\left. \left. \times \left[\left(x + \frac{h}{2} \right)^2 + y^2 \right] - \frac{1}{2l_1}\left[(x + a)^2 + y^2 \right] \right\} \right),$$

$$(5.5\text{-}3\mathrm{b})$$

$$t_{p\,\mathrm{rel}_3} = \psi_{p1} \psi_{p3}^* = \left(t_{p\,\mathrm{rel}_1} \right)^*, \quad (5.5\text{-}3\mathrm{c})$$

$$t_{p\,\mathrm{rel}_4} = \psi_{p2} \psi_{p3}^* = \left(t_{p\,\mathrm{rel}_2} \right)^*. \quad (5.5\text{-}3\mathrm{d})$$

Suppose, now, that the hologram just constructed is illuminated with a reconstruction wave from a point source 4, as shown in Figure 5.10(b), and having a wave number $k_2 = (2\pi/\lambda_2)$. Then the complex field ψ_{p4} illuminating the hologram is

$$\psi_{p4} = A_4 \exp\left(-jk_2\left\{l_2 + \frac{1}{2l_2}\left[(x-b)^2 + y^2\right]\right\}\right), \quad (5.5\text{-}4)$$

where A_4 is a complex constant. We find the total field immediately behind the hologram by multiplying Eq. (5.5-4) with Eq. (5.5-2), relevant terms of which are

$$\psi_{pR\,\text{rel}_i} = \psi_{p4} t_{p\,\text{rel}_i}, \quad i = 1, 2, 3, 4, \quad (5.5\text{-}5)$$

where the $t_{p\,\text{rel}}$s are defined in Eqs. (5.5-3a)–(5.5-3d).

Consider, first, the contribution from $\psi_{pR\,\text{rel}_1}$ and $\psi_{pR\,\text{rel}_2}$. After propagation through a distance z behind the hologram, these fields are transformed according to the Fresnel diffraction formula. Note also that because both these fields are converging [see Eqs. (5.5-3a) and (5.5-3b)], they will contribute to real images. Explicitly, the fields can be written [using Eqs. (3.3-19a) and (3.3-19b)] as

$$\psi_{p\,\text{rel}_1} \propto \left[\exp\left(jk_1\left\{(R - l_1) + \frac{1}{2R}\left[\left(x - \frac{h}{2}\right)^2 + y^2\right]\right.\right.\right.$$

$$\left.\left. - \frac{1}{2l_1}\left[(x+a)^2 + y^2\right]\right\}\right)$$

$$\times \exp\left(-jk_2\left\{l_2 + \frac{1}{2l_2}\left[(x-b)^2 + y^2\right]\right\}\right)\right]$$

$$* \exp\left[-jk_2\left(z + \frac{x^2 + y^2}{2z}\right)\right], \quad (5.5\text{-}6a)$$

$$\psi_{p\,\text{rel}_2} \propto \left[\exp\left(jk_1\left\{(R + d - l_1) + \frac{1}{2(R+d)}\left[\left(x + \frac{h}{2}\right)^2 + y^2\right]\right.\right.\right.$$

$$\left.\left. - \frac{1}{2l_1}\left[(x+a)^2 + y^2\right]\right\}\right)$$

$$\times \exp\left(-jk_2\left\{l_2 + \frac{1}{2l_2}\left[(x-b)^2 + y^2\right]\right\}\right)\right]$$

$$* \exp\left[-jk_2\left(z + \frac{x^2 + y^2}{2z}\right)\right]. \quad (5.5\text{-}6b)$$

Recall now that the convolutions above formally demand that we perform integrations by rewriting the functions in convolution with new independent variables x', y' and $(x - x')$, $(y - y')$, respectively. As seen in Chapter 4 in connection with lenses, it may be possible to equate the coefficients of x'^2, y'^2, appearing in the exponentials, to zero, thus leaving only linear terms in x', y'. [See, specifically, the steps in going from Eq. (4.1-3) to Eq. (4.1-6) for details.] In fact, doing this for Eq. (5.5-6a) gives

$$\frac{k_1}{2R} - \frac{k_1}{2l_1} - \frac{k_2}{2l_2} - \frac{k_2}{2z_{r_1}} = 0, \qquad (5.5\text{-}7a)$$

where we have replaced z by z_{r_1}. From Eq. (5.5-7a), we can solve for z_{r_1} to get

$$z_{r_1} = \left[\frac{k_1}{k_2}\left(\frac{1}{R} - \frac{R}{l_1} \right) - \frac{1}{l_2} \right]^{-1} = \frac{\lambda_1 R l_1 l_2}{\lambda_2 l_1 l_2 - (\lambda_2 l_2 + \lambda_1 l_1)R},$$

$$(5.5\text{-}7b)$$

where we have replaced the ks by the corresponding λs. At this distance, from Eq (5.5-6a),

$$\psi_{p\,\text{rel}_1} \propto \int\int_{-\infty}^{\infty} \exp\left(j\left\{ \left[k_1\left\{ -\frac{h}{2R} - \frac{a}{l_1} \right\} + k_2\left\{ \frac{b}{l_2} + \frac{x}{z_{r_1}} \right\} \right] x' \right.\right.$$

$$\left.\left. + \frac{k_2 y}{z_{r_1}} y' \right\} \right) dx'\, dy'$$

$$\propto \delta\left[x + z_{r_1}\left(\frac{b}{l_2} - \frac{k_1}{2k_2}\frac{h}{R} - \frac{k_1}{k_2}\frac{a}{l_1} \right), y \right], \qquad (5.5\text{-}8)$$

which is a δ function shifted in the lateral direction and is the real image of the point object 1. A similar analysis of Eq. (5.5-6b) reveals that this is also responsible for a real image, expressible as

$$\psi_{p\,\text{rel}_2} \propto \delta\left[x + z_{r_2}\left(\frac{b}{l_2} + \frac{k_1}{2k_2}\frac{h}{R+d} - \frac{k_1}{k_2}\frac{a}{l_1} \right), y \right], \qquad (5.5\text{-}9)$$

with

$$z_{r_2} = \left[\frac{k_1}{k_2} \left(\frac{1}{R + d} - \frac{1}{l_1} \right) - \frac{1}{l_2} \right]^{-1}$$

$$= \frac{\lambda_1 (R + d) l_1 l_2}{\lambda_2 l_1 l_2 - (\lambda_2 l_2 + \lambda_1 l_1)(R + d)}. \quad (5.5\text{-}10)$$

Alternatively, we could find $\psi_{p \, \mathrm{rel}_2}$ and z_{r_2} by comparing Eq. (5.5-6b) with Eq. (5.5-6a) and noting that we only need to change R to $R + d$ and h to $-h$ to derive the former from the latter; hence, the same changes in Eqs. (5.5-7b) and (5.5-8) would readily yield Eqs. (5.5-10) and (5.5-9), respectively. This gives the real image of the point object 2. These two real images are shown in Figure 5.10(b).

We are now in a position to evaluate *lateral* and *longitudinal magnifications*. For instance, the longitudinal distance (along z) between the two images is $z_{r_2} - z_{r_1}$, so that the longitudinal magnification is

$$M_{\mathrm{long}}^r \triangleq \frac{z_{r_2} - z_{r_1}}{d} \quad (5.5\text{-}11\mathrm{a})$$

$$\cong \frac{\lambda_1 \lambda_2 (l_1 l_2)^2}{(\lambda_2 l_1 l_2 - \lambda_2 R l_2 - \lambda_2 R l_1)^2}, \quad (5.5\text{-}11\mathrm{b})$$

using Eqs. (5.5-7b) and (5.5-10) and assuming $R \gg d$. We evaluate the lateral distance (along x) between the two images by taking the difference between the locations of the two δ functions in Eqs. (5.5-8) and (5.5-9), so that the lateral magnification is

$$M_{\mathrm{lat}}^r \triangleq \frac{z_{r_2} \left(\dfrac{b}{l_2} + \dfrac{k_1}{2k_2} \dfrac{h}{R + d} - \dfrac{k_1}{k_2} \dfrac{a}{l_1} \right) - z_{r_1} \left(\dfrac{b}{l_2} \dfrac{k_1}{2k_2} \dfrac{h}{R} - \dfrac{k_1}{k_2} \dfrac{a}{l_1} \right)}{h}$$

$$\cong \frac{\left[\left(\dfrac{b}{l_2} - \dfrac{\lambda_2}{\lambda_1} \dfrac{a}{l_1} \right)(z_{r_2} - z_{r_1}) + \dfrac{1}{2}(z_{r_1} + z_{r_2}) \dfrac{\lambda_2}{\lambda_1} \dfrac{h}{R} \right]}{h}. \quad (5.5\text{-}12)$$

In order to make this magnification independent of h (the lateral separation between the objects), we must set

$$\frac{b}{l_2} - \frac{\lambda_2}{\lambda_1} \frac{a}{l_1} = 0,$$

implying

$$\frac{b}{a} = \frac{\lambda_2}{\lambda_1} \frac{l_2}{l_1}. \qquad (5.5\text{-}13)$$

Then, from Eq. (5.5-12),

$$M_{\text{lat}}^r \approx \frac{(z_{r_1} + z_{r_2})}{2R} \frac{\lambda_2}{\lambda_1}$$

$$\approx \frac{\lambda_2 l_1 l_2}{\lambda_2 l_1 l_2 - (\lambda_2 l_2 + \lambda_2 l_1)R}, \qquad (5.5\text{-}14)$$

assuming $R \gg d$. Note, upon comparing Eqs. (5.5-11b) and (5.5-14), that

$$M_{\text{long}}^r \simeq \frac{\lambda_1}{\lambda_2} (M_{\text{lat}}^r)^2. \qquad (5.5\text{-}15)$$

The locations of the virtual images, as well as the lateral and longitudinal magnifications, can be similarly calculated starting from Eqs. (5.5-3c) and (5.5-3d).

The preceding analysis demonstrates clearly how *three-dimensional* images are formed from a hologram. As seen, image distortion (due to the lateral magnification not being equal to the longitudinal magnification) may, in general, result unless special care is taken to alleviate the problems. The preceding analysis also shows that, in principle, a hologram can be recorded using a particular color and read by some other color. In fact, sometimes this proves to be an advantage, because, as can be readily checked from Eq. (5.5-15), the choice

$$M_{\text{lat}}^r = \frac{\lambda_2}{\lambda_1} \qquad (5.5\text{-}16a)$$

Optical Information Processing (II)

ensures

$$M_{\text{lat}}^r = M_{\text{long}}^r, \qquad (5.5\text{-}16b)$$

thus eliminating any image distortion.

5.6 Holographic Resolution

In general, *holographic resolution* is limited by the physical size of the hologram, the aberrations in wavefront reconstruction, and the temporal coherence of the illuminating beam, to name a few. In what follows, we briefly discuss the effect of the physical size of the hologram on the nature of the reconstructed image.

Recall that a hologram illuminated by the reading beam gives, for instance, a real image of the object. In connection with the discussion in the preceding section, suppose that we now consider the recording and reconstruction of *one* point object, for example, the point 1 in Figure 5.10(a). Note that if $l_1 = l_2 = l$ and $\lambda_1 = \lambda_2$, the real image forms at a distance (see Eq. (5.5-7b))

$$z_{r_1} = \frac{Rl}{l - 2R} \qquad (5.6\text{-}1)$$

behind the plane of the hologram. We can ask the question: Could we replace the two-step process of recording and reconstruction by an equivalent single-lens imaging system? The answer is yes, provided we choose the focal length of the equivalent lens to be

$$\frac{1}{f} = \frac{1}{R} + \frac{1}{z_{r_1}}, \qquad (5.6\text{-}2a)$$

implying, with Eq. (5.6-1), that

$$f = \frac{Rl}{2(l - R)}. \qquad (5.6\text{-}2b)$$

This is, however, not the only criterion to be taken into account. Recall that the position of the real image in a single-lens imaging system is also fixed once we know the position of the object

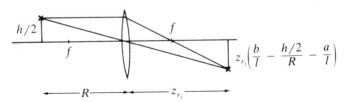

Figure 5.11 Equivalent single-lens imaging system model for holographic recording and reconstruction.

and the image distance. A simple geometrical construction shows that (see Figure 5.11)

$$\frac{z_{r_1}\left(\dfrac{b}{l} - \dfrac{h/2}{R} - \dfrac{a}{l}\right)}{h/2} = \frac{z_{r_1}}{R}, \qquad (5.6\text{-}3a)$$

where the quantity in the numerator on the LHS gives the location of the image, assuming $k_1 = k_2$ [see Eq. (5.5-8)], and where $h/2$ in the denominator on the LHS is the location of the object. Equation (5.6-3a) translates to the condition

$$\frac{b - a}{l} = \frac{h}{R} \qquad (5.6\text{-}3b)$$

and specifies the location of the recording and reconstruction sources necessary for this simple analogy. In other words, if the recording and reconstruction were performed abiding by the condition (5.6-3b), the two-step recording and reconstruction process may be depicted in terms of an equivalent single-lens imaging system where the object is placed a distance R in front of a lens having a focal length f given by Eq. (5.6-2b).

Now, in the previous chapter, readers were exposed to the impulse response of an imaging system. Recall that if $p_f(x, y)$ represents the pupil function of the lens, the impulse response is given by Eq. (4.2-13). If our hologram is of finite size, it can be conceived as a hologram of infinite dimensions preceded by or followed by a pupil function $p_f(x, y)$, and the impulse response

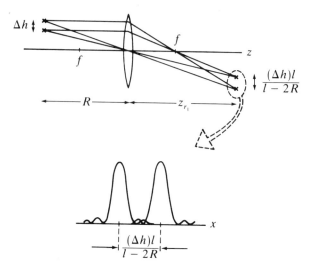

Figure 5.12 Definition of the resolution limit of a hologram.

would be

$$\mathscr{h}(x, y) = P_f\left(\frac{k_0}{z_{r_1}}x, \frac{k_0}{z_{r_1}}y\right), \qquad (5.6\text{-}4)$$

where z_{r_1} is given in Eq. (5.6-1), and the transfer function $\mathscr{H}(k_x, k_y)$ can be defined analogously to Eq. (4.2-14).

Suppose, now, that we have two points 1 and 1' close to each other as shown in Figure 5.12, with a distance Δh separating them on the transverse plane. Then, on the image plane, we will see their *smeared* images, given by Eq. (5.6-4), which are separated by an amount $(\Delta h)z_{r_1}/R = (\Delta h)l/(l - 2R)$. Whether we can distinguish the two images will be determined by the amount of overlap between the smeared images and can be stated in terms of the Rayleigh criterion (see Section 3.5), if the pupil function is

$$p_f(x, y) \triangleq \text{circ}\left(\frac{r}{r_0}\right), \qquad (5.6\text{-}5)$$

which then enunciates the resolution limit.

The simplistic analysis advanced here can be reworked formally for a more general case; however, the idea behind this presentation is to appreciate the analogy with a single-lens system.

Problems

5.1 A transparency $t(x, y)$ as shown in Figure 4.12 is placed on the object plane of an image processing system like that shown in Figure 4.7, with $f_1 = f_2 = f$, and illuminated with a plane wave. A photographic plate is placed on the image plane and a negative transparency t_n of $\gamma_{n1} = +2$ is produced. Sketch the amplitude transmittance of this transparency.

This transparency is thereafter reinserted on the object plane of the same image-processing system and illuminated as before. Sketch the intensity distribution on the image plane. Assume that there is a small opaque stop on the axis $(x = 0, y = 0)$ at the common focal plane.

5.2 Two transparencies $t_1(x, y)$ and $t_2(x, y)$ are placed a distance f in front of a converging lens of focal length f, as shown in Figure P5.2. If a film is placed on the back focal plane and a positive transparency with a γ of 2 is constructed, find an expression for the amplitude transmittance of the transparency.

This transparency is reinserted on the front focal plane of the same system. Find the intensity distribution on the back focal plane. Throughout the problem, assume plane-wave illumination of the input transparencies.

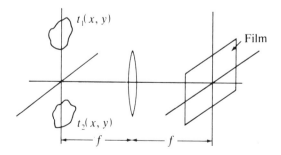

Figure P5.2

5.3   Assume that a Leith–Upatneiks hologram [as given by Eq. (5.3-2)] is placed on the front focal plane of a converging lens of focal length f. Find the intensity distribution at the back focal plane in terms of the original object $t(x, y)$ from which the hologram was constructed. Assume plane-wave illumination of the hologram.

5.4   With reference to Figure 5.5, assume that the object $t(x, y) = \delta(x - x_0, y - y_0)$. Find an expression for the positive transparency $t_p(x, y)$ constructed with an overall γ of 2.

This transparency t_p is illuminated by a plane wave travelling in the $+z$ direction as shown in Figure 5.7. Find the locations and nature of the real and virtual images of the object.

Suppose the reconstruction beam is, instead, a plane wave travelling at an angle θ_0 to the (z) axis. Predict the new locations of the real and virtual images.

5.5   A hologram constructed from Young's double-slit experiment, with an overall γ of 2, is illuminated with a reconstructing wave as shown in Figure P5.5. Find the locations of the real images of the two slits.

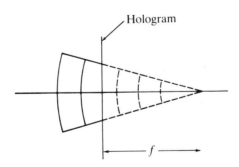

Figure P5.5

5.6   Develop expressions for the longitudinal and lateral magnifications for the virtual images due to the recording and reconstruction geometry as shown in Figure 5.10. Start by examining Eq. (5.5-3c) and (5.5-3d).

5.7 A hologram of a point source using a plane wave reference is first constructed with an overall γ of 2, as shown in Figure 5.4(a). Now a point object is located at a distance d_0 in front of the hologram. Find the location of the real image of the object.

References
5.1 Gabor, D. (1948). *Nature* **161** 777.
5.2 Ghatak, A. K. and K. Thyagarajan (1978). *Contemporary Optics*. Plenum, New York.
5.3 Goodman, J. W. (1968). *Introduction to Fourier Optics*. McGraw-Hill, New York.
5.4 Leith, E. N. and J. Upatneiks (1962). *J. Opt. Soc. Amer.* **52** 1123.
5.5 Lohmann, A. (1956). *Opt. Acta* **3** 97.
5.6 Nicodemus, F. E. (1983). Radiometry. In *Applied Optics and Optical Engineering* Vol. 9 (R. R. Shannon and J. C. Wyant, eds.) Chapter 13. Academic, New York.
5.7 Stroke, G. W. (1966). *An Introduction to Coherent Optics and Holography*. Academic Press, New York.
5.8 Yu, F. T. S. (1983). *Optical Information Processing*. Wiley, New York.

Chapter 6

Acoustooptics and Electrooptics

The ability to modulate light waves by electrical signals through either the acoustooptic or electrooptic effect provides a powerful means for optically processing information. In fact, some of the key components in modern optical processors usually consist of a *spatial light modulator*, such as an acoustooptic or electrooptic modulator, which is capable of spatially modulating the light beam. In this chapter, we deal with the acoustooptic and electrooptic effects. We shall also include some applications of such devices in signal processing. Sections 6.1–6.3 discuss the rudiments of acoustooptics, the mathematical formalism and selected applications. In Section 6.4, we introduce the topic of wave propagation in an anisotropic medium, as background for the electrooptic effect and its applications, discussed in Sections 6.5 and 6.6, respectively.

6.1 Acoustooptics: Qualitative Description and Heuristic Background

The interaction between sound and light is usually termed *acoustooptic interaction*. An *acoustooptic modulator* comprises an acoustic medium (such as glass or water) to which a piezoelectric transducer is bonded. Through the action of the piezoelectric transducer, the electrical signal is converted into sound waves propagating in the acoustic medium with an acoustic frequency spectrum that matches, within the bandwidth limitations of the transducer, that of the electrical excitation. The pressure in the sound wave creates a travelling wave of rarefaction and compression, which in turn causes analogous perturbations of the index of refraction. Thus, the acoustooptic device shown in Figure 6.1 may be thought to act as a thin (phase) grating with an effective grating line separation equal to the wavelength Λ of the sound in the acoustic medium. It is well known that a phase grating splits incident light into various orders (see Problem 3.19). It can be shown that the directions of the diffracted, or scattered, light inside the sound cell are governed by the *grating equation*,

$$\sin \phi_m = \sin \phi_{\text{inc}} + m\lambda_0/\Lambda, \qquad m = 0, \pm 1, \pm 2, \ldots, \quad (6.1\text{-}1)$$

where ϕ_m is the angle of the mth-order scattered light beam, ϕ_{inc} is the angle of incidence, and λ_0 is the wavelength of the light, all in the acoustic medium. Hence, the angle between neighboring orders in Figure 6.1 equals λ_0/Λ in the cell. When measured outside the

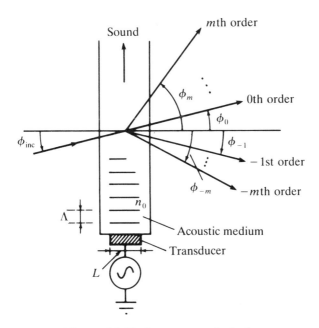

Figure 6.1 Basic acoustooptic device.

Acoustooptics and Electrooptics

medium, these angles are increased through refraction and can be found by multiplying Eq. (6.1-1) by the refractive index n_0 of the material of the sound cell. In all figures in this chapter, we assume all pertinent angles to be measured *inside* the sound cell. Because we really have a travelling sound wave, the frequencies of the scattered beams (except the 0th-order beam) in Figure 6.1 are either downshifted or upshifted by an amount equal to the sound frequency due to the *Doppler effect*, as will be seen later.

Through the relationship between the sound wavelength Λ and the sound velocity V_s in the acoustic medium ($\Lambda = 2\pi V_s/\Omega$), we can readily see that by electronically varying the sound frequency Ω, we can change the directions of propagation of the scattered beams. It is this feature that makes it possible to use an acoustooptic device as a spectrum analyzer, as we shall see in a later section. The frequencies of the sound waves produced in laboratories range from about 100 kHz to 3 GHz, and these sound waves are really ultrasound waves whose frequencies are not audible to the human ear. The range of the sound velocity in the medium lies from about 1 km/s in water to about 6.5 km/s in a crystalline material such as $LiNbO_3$.

As it turns out, the phase-grating treatment of acoustooptic interaction is somewhat of an oversimplification in that the approach does not predict the required angle for incident light in order to obtain efficient operation, nor does it explain why only one order is generated for a sufficiently wide transducer (L large in Figure 6.1).

Another, more accurate, approach considers the interaction of sound and light as a collision of photons and phonons. For these particles to have well-defined momenta and energies, we must assume that, classically, we have interaction of monochromatic plane waves of light and sound, that is, we assume that the width L of the transducer is sufficiently wide in order to produce plane wave fronts at a single frequency. In the process of collision, two conservation laws have to be obeyed, namely, the *conservation of energy and momentum*. If we denote the propagation vectors (also called *wave vectors*) of incident light, scattered light, and sound in the acoustic medium by $\mathbf{k}_0$, $\mathbf{k}_{+1}$, and $\mathbf{K}$, respectively, as shown in Figure 6.2, we can write the condition for conservation of momentum as

$$\hbar\mathbf{k}_{+1} = \hbar\mathbf{k}_0 + \hbar\mathbf{K}, \qquad (6.1\text{-}2)$$

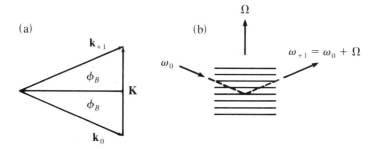

Figure 6.2 Upshifted diffraction: (a) wave-vector diagram; (b) experimental configuration.

where $\hbar = h/2\pi$ and $h = 6.63 \times 10^{-34}$(J-s) denotes Planck's constant. Dividing Eq. (6.1-2) by $\hbar$ leads to

$$\mathbf{k}_{+1} = \mathbf{k}_0 + \mathbf{K}. \qquad (6.1\text{-}3)$$

The corresponding conservation of energy takes the form (after division by $\hbar$)

$$\omega_{+1} = \omega_0 + \Omega, \qquad (6.1\text{-}4)$$

where ω_0, Ω, and ω_{+1} are the (radian) frequencies of the incident light, sound, and scattered light, respectively. The interaction described by Eq. (6.1-3) is called the *upshifted interaction*. Figure 6.2(a) shows the *wave-vector diagram*, and Figure 6.2(b) describes the diffracted beam being upshifted in frequency. Because for all practical cases $|\mathbf{K}| \ll |\mathbf{k}_0|$, the magnitude of $\mathbf{k}_{+1}$ is essentially equal to $\mathbf{k}_0$, and therefore the wave-vector momentum triangle shown in Figure 6.2(a) is nearly isosceles.

Suppose now that we change the directions of the incident and scattered light as shown in Figure 6.3. The conservation laws can be applied again to obtain two equations similar to Eqs. (6.1-3) and (6.1-4). The two equations describing the interaction are now

$$\mathbf{k}_{-1} = \mathbf{k}_0 - \mathbf{K} \qquad (6.1\text{-}5)$$

Acoustooptics and Electrooptics

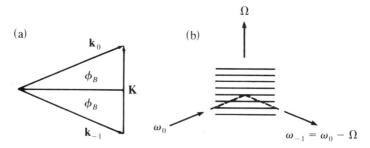

Figure 6.3 Downshifted diffraction: (a) wave-vector diagram; (b) experimental configuration.

and

$$\omega_{-1} = \omega_0 - \Omega, \qquad (6.1\text{-}6)$$

where the subscripts on the LHS indicate that the interaction is *downshifted*.

There is some interesting physics hidden in Eqs. (6.1-3)–(6.1-6). It can be shown that Eqs. (6.1-3) and (6.1-4) refer to *phonon absorption* and Eqs. (6.1-5) and (6.1-6) to *stimulated phonon emission*. Indeed, attenuation and amplification of a sound wave have been demonstrated for these cases [see Korpel, Adler, and Alpiner (1964)].

We have seen that the wave-vector diagrams [Figures 6.2(a) and 6.3(a)] must be closed for both cases of interaction. The closed diagrams stipulate that there are certain critical angles of incidence $(\pm\phi_B)$ in the acoustic medium for plane waves of sound and light to interact, and also that the directions of the incident and scattered light differ in angle by $2\phi_B$. The angle ϕ_B is called the *Bragg angle*, and this form of diffraction is called *Bragg diffraction*, analogous to X-ray diffraction in crystals. From Figure 6.2(a) or 6.3(a), the Bragg angle is given by

$$\sin \phi_B = \frac{K}{2k_0} = \frac{\lambda_0}{2\Lambda}. \qquad (6.1\text{-}7)$$

Acoustooptics

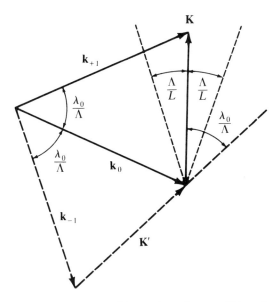

Figure 6.4 Wave-vector diagram illustrating the condition for defining the Bragg regime.

In actuality, scattering occurs even though the direction of incident light is not exactly at the Bragg angle. However, the maximum scattered intensity occurs at the Bragg angle. The reason is that we do not have exactly plane wavefronts; the sound waves actually spread out as they propagate into the medium. As the width L of the transducer decreases, the sound column will act less and less like a single plane wave and, in fact, it is now more appropriate to consider an angular spectrum of plane waves. For a transducer with an aperture L, sound waves spread out over an angle $\pm \Lambda/L$ if we use the heuristic approach to propagational diffraction (see Chapter 3). Considering the upshifted Bragg interaction and, referring to Figure 6.4, we see that the **K**-vector can be oriented through an angle $\pm \Lambda/L$ due to the spreading of sound. In order to have only one diffracted order of light generated (i.e., $\mathbf{k}_{+1}$), we have to impose the condition

$$\frac{\lambda_0}{\Lambda} \gg \frac{\Lambda}{L},$$

Acoustooptics and Electrooptics

or

$$L \gg \frac{\Lambda^2}{\lambda_0}. \qquad (6.1\text{-}8)$$

This is because for $\mathbf{k}_{-1}$ to be generated, for example, a pertinent sound wave vector must lie along $\mathbf{K}'$; however, this either is not present, or is present in negligible amounts in the angular spectrum of the sound, if the condition (6.1-8) is satisfied. If L satisfies the condition (6.1-8), the acoustooptic device is said to operate in the *Bragg regime* and the device is commonly known as the Bragg cell.

In the case where L is sufficiently short, we have the second form of scattering (diffraction), called *Raman–Nath* (or *Debye–Sears*) *diffraction*. The condition

$$L \ll \frac{\Lambda^2}{\lambda_0}, \qquad (6.1\text{-}9)$$

therefore, defines the *Raman–Nath diffraction regime*.

In Raman–Nath diffraction, $\mathbf{k}_{+1}$ and $\mathbf{k}_{-1}$ (i.e., positive and negative first order scattered light) are generated simultaneously because various directions of plane waves of sound are provided from a small-aperture transducer. So far we have only considered the so-called *weak interaction* between the sound and incident light, that is, the interaction between the scattered light and sound has been ignored. In fact, scattered light may interact with the sound field again and produce higher orders of diffracted light. This rescattering process is characteristic of what is called *strong interaction*. (In the Bragg regime, for strong interaction, scattered light $\mathbf{k}_{+1}$ may rescatter back into the zero-order light). In the Raman–Nath regime, many orders may exist because plane waves of sound are available at the various angles required for scattering. The principle of the generation of many orders by rescattering is illustrated in Figure 6.5: $\mathbf{k}_{+1}$ is generated through the diffraction of $\mathbf{k}_0$ by $\mathbf{K}_{+1}$, $\mathbf{k}_{+2}$ is generated through the diffraction of $\mathbf{k}_{+1}$ by $\mathbf{K}_{+2}$, and so on, where the $\mathbf{K}_{\pm p}$'s denote the appropriate components of the plane-wave spectrum of the sound. Again, the requirement of conserva-

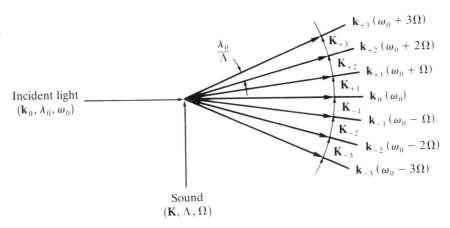

Figure 6.5 Multiple scattering in the Raman–Nath regime.

tion of energy leads to the equation $\omega_m = \omega_0 \pm m\Omega$, with ω_m being the frequency of the mth-order scattered light.

6.2 The Acoustooptic Effect: General Formalism

The interaction between the optical field $\mathbf{E}_0(\mathbf{r}, t)$ and sound field $S(\mathbf{r}, t)$ can be generally described by Maxwell's equations, Eqs. (3.1-1)–(3.1-4). We assume that the interaction takes place in an optically inhomogeneous, nonmagnetic isotropic medium, characterized by a permeability μ_0 and a permittivity $\tilde{\epsilon}(\mathbf{r}, t)$ when a source-free optical field is incident on the time-varying permittivity. The time-varying permittivity is written as

$$\tilde{\epsilon}(\mathbf{r}, t) = \epsilon + \epsilon'(\mathbf{r}, t), \qquad (6.2\text{-}1)$$

where $\epsilon'(\mathbf{r}, t) = \epsilon CS(\mathbf{r}, t)$, that is, it is proportional to the sound field amplitude $S(\mathbf{r}, t)$, with C the proportionality constant, dependent on the medium. Hence $\epsilon'(\mathbf{r}, t)$ represents the action of the sound field. The analysis presented below closely follows the work by Korpel (1972).

We will assume $\mathbf{E}_{inc}(\mathbf{r}, t)$ satisfies Maxwell's equations [Eqs. (3.1-1)–(3.1-4)], with $\rho = 0$ and $\mathbf{J}_c = 0$. When the sound field interacts with $\mathbf{E}_{inc}(\mathbf{r}, t)$, the total field $\mathbf{E}(\mathbf{r}, t)$ in the cell must also satisfy

Acoustooptics and Electrooptics

Maxwell's equations, rewritten as

$$\nabla \times \mathbf{E}(\mathbf{R}, t) = -\mu_0 \frac{\partial \mathbf{H}(\mathbf{R}, t)}{\partial t}, \qquad (6.2\text{-}2)$$

$$\nabla \times \mathbf{H}(\mathbf{R}, t) = \frac{\partial}{\partial t}[\tilde{\epsilon}(\mathbf{R}, t)\mathbf{E}(\mathbf{R}, t)], \qquad (6.2\text{-}3)$$

$$\nabla \cdot [\tilde{\epsilon}(\mathbf{R}, t)\mathbf{E}(\mathbf{R}, t)] = 0, \qquad (6.2\text{-}4)$$

$$\nabla \cdot \mathbf{H}(\mathbf{R}, t) = 0. \qquad (6.2\text{-}5)$$

Taking the curl of Eq. (6.2-2) and introducing it into Eq. (6.2-3), the equation for $\mathbf{E}(\mathbf{R}, t)$ reads

$$\nabla \times \nabla \times \mathbf{E}(\mathbf{R}, t) = \nabla(\nabla \cdot \mathbf{E}) - \nabla^2 \mathbf{E}$$

$$= -\mu_0 \frac{\partial^2}{\partial t^2}[\tilde{\epsilon}(\mathbf{R}, t)\mathbf{E}(\mathbf{R}, t)]. \qquad (6.2\text{-}6)$$

Now,

$$\nabla \cdot \tilde{\epsilon}\mathbf{E} = \tilde{\epsilon}\nabla \cdot \mathbf{E} + \mathbf{E} \cdot \nabla\tilde{\epsilon} = 0, \qquad (6.2\text{-}7)$$

using Eq. (6.2-4). Assuming a two-dimensional $(x\text{--}z)$ configuration with $\mathbf{E}$ polarized along the y direction, we can readily show that $\mathbf{E} \cdot \nabla\tilde{\epsilon} = 0$; hence Eq. (6.2-6) reduces to

$$\nabla^2 E(\boldsymbol{\rho}, t) = \mu_0 \frac{\partial^2}{\partial t^2}[\tilde{\epsilon}(\boldsymbol{\rho}, t)E(\boldsymbol{\rho}, t)], \qquad (6.2\text{-}8)$$

where $\boldsymbol{\rho}$ is the position vector in the $x\text{--}z$ plane and $\mathbf{E}(\boldsymbol{\rho}, t) = E(\boldsymbol{\rho}, t)\hat{a}_y$. We will rewrite the term on the right-hand side of Eq. (6.2-8) as

$$\mu_0 \left[E \frac{\partial^2 \tilde{\epsilon}}{\partial t^2} + 2 \frac{\partial E}{\partial t} \frac{\partial \tilde{\epsilon}}{\partial t} + \tilde{\epsilon} \frac{\partial^2 E}{\partial t^2} \right]. \qquad (6.2\text{-}9)$$

Because the time variation of $\tilde{\epsilon}(\boldsymbol{\rho}, t)$ is much slower than that of $E(\boldsymbol{\rho}, t)$, we will only retain the last term in Eq. (6.2-9) to get, using

Eqs. (6.2-1) and (6.2-8),

$$\nabla^2 E(\boldsymbol{\rho}, t) - \mu_0 \epsilon \frac{\partial^2 E(\boldsymbol{\rho}, t)}{\partial t^2} = \mu_0 \epsilon'(\boldsymbol{\rho}, t) \frac{\partial^2 E(\boldsymbol{\rho}, t)}{\partial t^2}, \quad (6.2\text{-}10)$$

Equation (6.2-10) is the wave equation that is often used to investigate strong interaction in acoustooptics.

We will now introduce harmonic variations in the incident light and sound in the forms

$$E_{\text{inc}}(\boldsymbol{\rho}, t) = \frac{1}{2} E_{\text{inc}_p}(\boldsymbol{\rho}) e^{j\omega_0 t} + \text{c.c.}, \quad (6.2\text{-}11)$$

and

$$\frac{\epsilon'(\boldsymbol{\rho}, t)}{\epsilon} = \frac{1}{2} CS_p(\boldsymbol{\rho}) e^{j\Omega t} + \text{c.c.}, \quad (6.2\text{-}12)$$

where c.c. denotes the complex conjugate, and the subscript p reminds us that the corresponding variables denote phasors. For notational convenience, we will drop the subscript p in subsequent discussions. Because we have harmonic fields, we anticipate, as seen in the simplified analysis before in Section 6.1, frequency mixing due to the interaction. This is also evident from examining the RHS of Eq. (6.2-10). We will therefore cast the total field $E(\boldsymbol{\rho}, t)$ into the form

$$E(\boldsymbol{\rho}, t) = \frac{1}{2} \sum_{m=-\infty}^{\infty} E_m(\boldsymbol{\rho}) \exp\left[j(\omega_0 + m\Omega)t \right] + \text{c.c.} \quad (6.2\text{-}13)$$

Substituting Eqs. (6.2-11)–(6.2-13) into Eq. (6.2-10) and assuming $\Omega \ll \omega_0$, we obtain, after some straightforward calculations, the following infinite coupled-wave system:

$$\nabla^2 E_m(\boldsymbol{\rho}) + k_0^2 E_m(\boldsymbol{\rho}) + \frac{1}{2} k_0^2 CS(\boldsymbol{\rho}) E_{m-1}(\boldsymbol{\rho})$$

$$+ \frac{1}{2} k_0^2 CS^*(\boldsymbol{\rho}) E_{m+1}(\boldsymbol{\rho}) = 0, \quad (6.2\text{-}14)$$

where $k_0 = \omega_0 \sqrt{\mu_0 \epsilon}$ is the propagation constant of light in the

Acoustooptics and Electrooptics

medium and the asterisk denotes the complex conjugate. Note that $E_m(\boldsymbol{\rho})$ is the phasor amplitude of the mth-order light at frequency $\omega_0 + m\Omega$.

6.2.1 Sound-Field Configuration and Plane-Wave Interaction Model

We will now consider a conventional interaction configuration depicted in Figure 6.6. As we can see, a uniform sound wave of finite width L and propagating along x is often used, and we shall represent the sound by

$$S(\boldsymbol{\rho}) = S(z, x) = A e^{-jKx}, \qquad (6.2\text{-}15)$$

where A, in general, can be complex. The incident plane wave of light will be represented as

$$E_{\text{inc}}(\boldsymbol{\rho}) = \psi_{\text{inc}} \exp(-jk_0 z \cos \phi_{\text{inc}} - jk_0 x \sin \phi_{\text{inc}}), \quad (6.2\text{-}16)$$

where ϕ_{inc} is the incident angle. We look for a solution of $E_m(\boldsymbol{\rho})$ of

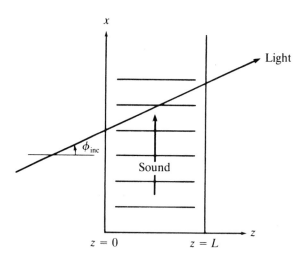

Figure 6.6 Conventional sound–light interaction configuration.

the form

$$E_m(\mathbf{\rho}) = E_m(z, x) = \psi_m(x, z)\exp(-jk_0 z \cos \phi_m - jk_0 x \sin \phi_m),$$

$$(6.2\text{-}17)$$

with the choice for ϕ_m given by Eq. (6.1-1),

$$\sin \phi_m = \sin \phi_{\text{inc}} + m\frac{\lambda_0}{\Lambda} = \sin \phi_{\text{inc}} + m\frac{K}{k_0}. \qquad (6.2\text{-}18)$$

Substituting Eqs. (6.2-15), (6.2-17), and (6.2-18) into Eq. (6.2-14), we obtain, after some algebra,

$$\frac{\partial^2 \psi_m}{\partial x^2} - 2jk_0 \sin \phi_m \frac{\partial \psi_m}{\partial x} - 2jk_0 \cos \phi_m \frac{\partial \psi_m}{\partial z}$$

$$+ \frac{1}{2}k_0^2 CA^* \psi_{m+1} \exp\left[-jk_0 z(\cos \phi_{m+1} - \cos \phi_m)\right]$$

$$+ \frac{1}{2}k_0^2 CA \psi_{m-1} \exp\left[-jk_0 z(\cos \phi_{m-1} - \cos \phi_m)\right] = 0. \quad (6.2\text{-}19)$$

In deriving Eq. (6.2-19), we have assumed that, within a wavelength of light, $\partial \psi_m/\partial z$ does not change appreciably; hence, $\partial^2 \psi_m/\partial z^2$ can be neglected when compared to $(2jk_0 \cos \phi_m)\partial \psi_m/\partial z$. The physics behind Eq. (6.2-19) can be understood as follows: The evolution of the mth-order scattered light in z depends on (a) interactions between adjacent orders ($m \pm 1$) with the sound (the last two terms on the LHS) and (b) the effect of propagational diffraction (the first term on the LHS). Indeed, the latter is identical to that in Eq. (3.3-11), which was argued to be responsible for diffraction. The second term on the LHS is merely the effect of the mth-order light travelling in a direction slightly different from z.

Often, the effect of propagational diffraction can be neglected if the width L of the cell is not too large. Also, because ϕ_m is a very small quantity (if $\phi_{\text{inc}} \ll 1$), we can assume $\psi_m(z, x) \approx$

$\psi_m(z)$. Hence, Eq. (6.2-19) becomes

$$\frac{d\psi_m}{dz} = -j\frac{k_0 CA}{4\cos\phi_m}\psi_{m-1}\exp[-jk_0 z(\cos\phi_{m-1} - \cos\phi_m)]$$

$$- j\frac{k_0 CA^*}{4\cos\phi_m}\psi_{m+1}\exp[-jk_0 z(\cos\phi_{m+1} - \cos\phi_m)], \quad (6.2\text{-}20)$$

which has to be solved with the boundary conditions

$$\psi_m = \psi_{\text{inc}}\delta_{m0} \quad \text{at } z \le 0, \tag{6.2-21}$$

where δ_{m0} represents the Kronecker delta function.

The physical interpretation of Eq. (6.2-20), as stated before, is that there is a mutual coupling between neighboring orders in the interaction, that is, ψ_m is being contributed to by ψ_{m-1} and ψ_{m+1}. However, the phase of the contributions varies with z, and the exponents with arguments $k_0 z(\cos\phi_{m-1} - \cos\phi_m)$ and $k_0 z$ $(\cos\phi_{m+1} - \cos\phi_m)$ represent the lack of phase synchronism in this coupling process.

6.2.2 Raman – Nath Regime

As mentioned before, Raman–Nath diffraction is character-ized by the simultaneous generation of many scattered orders. This implies that the interaction length L must be short enough such that the accumulated degree of phase mismatch between ψ_m and its neighboring orders, as given by the arguments of the exponential terms in Eq. (6.2-20), is small. Assuming small values of K/k_0, we expand the phase-asynchronism terms $k_0 z(\cos\phi_{m-1} - \cos\phi_m)$ and $k_0 z(\cos\phi_{m+1} - \cos\phi_m)$ in Eq. (6.2-20) in a power series by using Eq. (6.2-18),

$$k_0 z(\cos\phi_{m-1} - \cos\phi_m)$$

$$= k_0 z\left[\left(\frac{K}{k_0}\right)\sin\phi_{\text{inc}} + \left(m - \frac{1}{2}\right)\left(\frac{K}{k_0}\right)^2 + \cdots\right], \tag{6.2-22}$$

$$k_0 z(\cos\phi_{m+1} - \cos\phi_m)$$

$$= k_0 z\left[-\left(\frac{K}{k_0}\right)\sin\phi_{\text{inc}} - \left(m + \frac{1}{2}\right)\left(\frac{K}{k_0}\right)^2 + \cdots\right]. \tag{6.2-23}$$

For $\phi_{inc} = 0$, their accumulated phase mismatch for order m at $z = L$ is negligible if

$$m\left(\frac{K^2}{k_0}\right)L \ll 1. \tag{6.2-24}$$

This condition is the criterion for an acoustooptic device operated in the Raman–Nath regime. It is usually stated with $m = 1$ and is consistent with Eq. (6.1-9).

Using this criterion and assuming $\phi_m \ll 1$, Eq. (6.2-20) becomes

$$\frac{d\psi_m}{dz} = -\frac{jk_0CA}{4}\psi_{m-1} - \frac{jk_0CA^*}{4}\psi_{m+1}. \tag{6.2-25}$$

For simplicity, let $A = A^* = |A|$, where $|A|$ represents the peak strain caused by the sound field $|A|\cos(\Omega t - Kx)$ [see Eqs. (6.2-12) and (6.2-15)]. Thus, from Eq. (6.2-25),

$$\frac{d\psi_m}{dz} = \frac{-jk_0C|A|}{4}[\psi_{m-1} + \psi_{m+1}]. \tag{6.2-26}$$

Now recall the recursion relation for the Bessel functions (the J_m's),

$$\frac{dJ_m(z)}{dz} = \frac{1}{2}[J_{m-1}(z) - J_{m+1}(z)]. \tag{6.2-27}$$

Then, writing $\psi_m = (-j)^m\psi_m'$, where $\psi_m' = J_m(k_0C|A|z/2)$, we recognize, with Eq. (6.2-21), that the amplitude of the various scattered orders at $z = L$ is

$$\psi_m = (-j)^m\psi_{inc}J_m\left(\frac{k_0C|A|L}{2}\right). \tag{6.2-28}$$

Equation (6.2-28) is the well-known Raman–Nath solution.

6.2.3 Bragg Regime

As pointed out before, Bragg diffraction is characterized by the generation of two scattered orders. Referring to Eq. (6.2-20), we find that for the diffracted orders 0 and -1 (downshifted interaction), the coupled equations read

$$\frac{d\psi_0}{dz} = \frac{-jk_0CA}{4\cos\phi_0}\psi_{-1}\exp\left[-jk_0z(\cos\phi_{-1} - \cos\phi_0)\right] \quad (6.2\text{-}29a)$$

and

$$\frac{d\psi_{-1}}{dz} = \frac{-jk_0CA^*}{4\cos\phi_{-1}}\psi_0\exp\left[+jk_0z(\cos\phi_{-1} - \cos\phi_0)\right]. \quad (6.2\text{-}29b)$$

Assuming phase synchronism between the 0th and -1st orders, we must impose $\cos\phi_{-1} = \cos\phi_0$, implying

$$\phi_{-1} = -\phi_0, \quad (6.2\text{-}30)$$

as $\phi_{-1} \neq \phi_0$, because different scattered orders must exit at different angles. Hence, referring to Figure 6.7(a),

$$\phi_0 = \phi_B = \phi_{\text{inc}},$$

$$\phi_{-1} = -\phi_B = -\phi_{\text{inc}}. \quad (6.2\text{-}31)$$

Thus, the 0th and the diffracted (-1st) order propagate symmetrically with respect to the sound wavefronts. Similar arguments may be advanced for the upshifted interaction case.

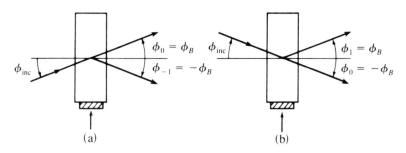

(a)　　　　　　　　　　(b)

Figure 6.7 (a) Downshifted interaction ($\phi_{\text{inc}} = +\phi_B$). (b) Upshifted interaction ($\phi_{\text{inc}} = -\phi_B$). The convention for angles is that counterclockwise is positive.

In light of the preceding discussion, the coupled equations describing the downshifted interaction become

$$\frac{d\psi_0}{dz} = \frac{-jk_0 CA}{4\cos\phi_B}\psi_{-1}, \tag{6.2-32}$$

$$\frac{d\psi_{-1}}{dz} = \frac{-jk_0 CA^*}{4\cos\phi_B}\psi_0. \tag{6.2-33}$$

Similarly, the coupled equations

$$\frac{d\psi_0}{dz} = \frac{-jk_0 CA^*}{4\cos\phi_B}\psi_1 \tag{6.2-34}$$

and

$$\frac{d\psi_1}{dz} = \frac{-jk_0 CA}{4\cos\phi_B}\psi_0 \tag{6.2-35}$$

describe the upshifted interaction ($\phi_{inc} = -K/2k_0 = -\phi_B$). Note that these equations are the standard ordinary differential equations (ODEs) for coupled modes, where the coupling coefficients are $-jk_0 CA/4\cos\phi_B$ and $-jk_0 CA^*/4\cos\phi_B$. The solutions at $z = L$, taking the boundary conditions Eq. (6.2-21) into account, and with $\phi_B \ll 1$, read

$$\psi_0 = \psi_{inc}\cos\left(\frac{K_0 C|A|L}{4}\right),$$

$$\psi_{-1} = -j\frac{A^*}{|A|}\psi_{inc}\sin\left(\frac{k_0 C|A|L}{4}\right), \tag{6.2-36}$$

for the downshifted interaction, and

$$\psi_0 = \psi_{inc}\cos\left(\frac{k_0 C|A|L}{4}\right),$$

$$\psi_1 = -j\frac{A}{|A|}\psi_{inc}\sin\left(\frac{k_0 C|A|L}{4}\right), \tag{6.2-37}$$

Acoustooptics and Electrooptics

for the upshifted case. Equations (6.2-36) or (6.2-37) are the well-known expressions for the scattered light in Bragg diffraction.

6.2.4 Discussion

At this point it is instructive to relate the term $C|A|$ to the refractive index variation $\Delta n(\mathbf{\rho}, t)$ in the acoustooptic cell. Because

$$\tilde{\epsilon}(\mathbf{\rho}, t) = \epsilon_0 n^2(\mathbf{\rho}, t)$$

$$= \epsilon_0 [n + \Delta n(\mathbf{\rho}, t)]^2$$

$$\simeq \epsilon_0 n^2 \left[1 + \frac{2 \Delta n(\mathbf{\rho}, t)}{n} \right], \tag{6.2-38}$$

where n denotes the unperturbed refractive index of the medium, we compare Eq. (6.2-38) with Eq. (6.2-1) and obtain

$$CS(\mathbf{\rho}, t) = \frac{2 \Delta n(\mathbf{\rho}, t)}{n}, \tag{6.2-39}$$

or $C|S| = C|A| = 2(\Delta n)_{\max}/n$, where $(\Delta n)_{\max}$ denotes the peak amplitude of the assumed harmonic variation of $\Delta n(\mathbf{\rho}, t)$. The quantity $k_0 C|A|L/2$ then can be written as

$$\frac{k_0 C|A|L}{2} = \left(\frac{k_0}{n} \right) (\Delta n)_{\max} L = \hat{\alpha}, \tag{6.2-40}$$

where $\hat{\alpha}$ represents the *peak phase delay* of the light through the acoustic medium.

Summarizing the solutions to the Raman–Nath and Bragg regions in terms of $\hat{\alpha}$, we have

$$\psi_m(z) = \psi_{\text{inc}}(-j)^m J_m(\hat{\alpha}), \tag{6.2-41}$$

and

$$\psi_0 = \psi_{\text{inc}} \cos(\hat{\alpha}/2),$$
$$\psi_{\pm 1} = -j\psi_{\text{inc}} \sin(\hat{\alpha}/2), \tag{6.2-42}$$

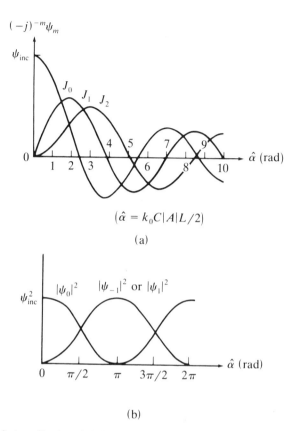

$$(\hat{\alpha} = k_0 C|A|L/2)$$

(a)

(b)

Figure 6.8 (a) Amplitude of diffracted orders in the Raman–Nath regime. (b) Intensity of diffracted orders versus peak phase delay $\hat{\alpha}$ in the Bragg region.

respectively, where we have assumed $A = A^* = |A|$. Figure 6.8 illustrates the dependence of various scattered orders on $\hat{\alpha}$ for the two regimes.

Finally, it should be pointed out that the criterion for an acoustooptic cell to operate in the Bragg regime is the inverse relation to Eq. (6.2-24),

$$(K^2/k_0)L \gg 1, \qquad (6.2\text{-}43)$$

where the strongest condition is implied (i.e., $m = 1$). [Note that

Acoustooptics and Electrooptics

Eq. (6.2-43) is consistent with Eq. (6.1-8).] In physical reality, a complete energy transfer between ψ_0 and $\psi_{\pm 1}$ is never possible, as there always exist more than two orders no matter how strong condition (6.2-43) becomes. This regime is known as the *near Bragg regime* [Poon and Korpel (1981)] and $|\psi_0|^2 + |\psi_{\pm 1}|^2 \neq \psi_{inc}^2$ due to the generation of higher orders. To establish the Bragg region more precisely, the so-called *Klein–Cook parameter Q* has been defined as

$$Q = (K^2/k_0)L, \qquad (6.2\text{-}44)$$

by Korpel (1972, 1988). The amount of first-order scattered light can be plotted as a function of Q at $\hat{\alpha} = \pi$. The Bragg regime is then defined arbitrarily by the condition that $|\psi_{\pm 1}|^2 > 0.9\psi_{inc}^2$ (i.e., the *diffraction efficiency* for the first-order light is greater than 90%), which has been shown to translate to $Q = (K^2/k_0)L > 7$ [Korpel (1972)]. For $Q \to \infty$, $|\psi_{\pm 1}|^2 \to \psi_{inc}^2$ as expected.

6.3 Some Applications of the Acoustooptic Effect

6.3.1 Intensity Modulation of a Laser Beam

Figure 6.9(a) shows the schematic diagram of an *intensity-modulation* system. The acoustooptic modulator is operated in the Raman–Nath regime and only the zeroth-order diffracted beam is allowed to pass through the aperture stop. With reference to Figure 6.9(b), we can achieve linear operation with a bias voltage b corresponding to the peak phase shift of the light approximately equal to 1.3 rad. Figure 6.9(b) illustrates the relationship between the modulating signal $A(t)$ and the intensity-modulated output of the laser beam. Although in Fig. 6.9(a) the zeroth-order beam is used, it is clear that any diffracted order can be used in principle. In that case, the aperture stop must be positioned at that particular beam used to achieve the modulation. Also, it is clear that intensity modulation can be achieved similarly if the acoustooptic modulator is operated in the Bragg regime.

6.3.2 Light Beam Deflector and Spectrum Analyzer

In contrast to intensity modulation, where the amplitude of the modulating signal is varied, the frequency of the modulating signal is changed for applications in light deflection. Figure 6.10

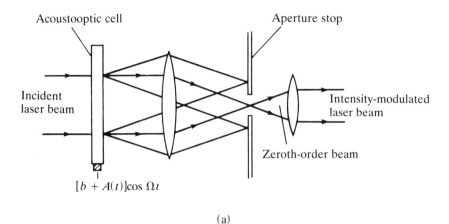

$[b + A(t)]\cos \Omega t$

(a)

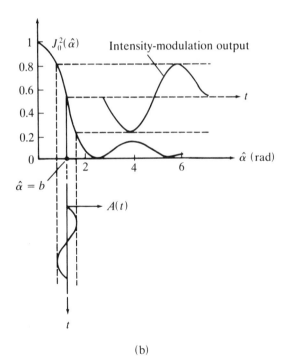

(b)

Figure 6.9 (a) An acoustooptic intensity modulation system. (b) Relationship between the modulating signal $A(t)$ and the intensity output.

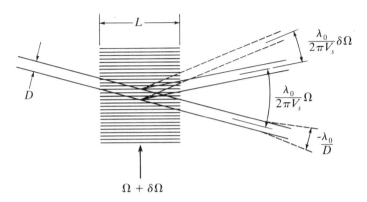

Figure 6.10 An acoustooptic light beam deflector.

shows a *light beam deflector* where the acoustooptic modulator is operated in the Bragg regime. The angle between the first-order beam and the zeroth-order beam is defined as the *deflection angle* ϕ_d. We can express the change in the deflection angle $\delta\phi_d$ upon a change $\delta\Omega$ of the sound frequency as

$$\delta\phi_d = \delta(2\phi_B)$$

$$= \frac{1}{2\pi}\frac{\lambda_0}{V_s}\delta\Omega. \tag{6.3-1}$$

Using a He–Ne laser ($\lambda_0 \sim 0.6 \ \mu$m), a change of sound frequency of 20 MHz around the center frequency of 40 MHz, and $V_s \sim 4 \times 10^3$ m/s for the velocity of sound in glass, a change in deflection angle is $\delta\phi_d \sim 3$ mrad.

The *number of resolvable angles N* in such a device is determined by the ratio of the range of deflection angles $\delta\phi_d$ to the angular spread of the scanning light beam. The angular spread of a beam of width D is of the order of λ_0/D, which follows from standard diffraction theory (see Chapter 3). Hence,

$$N = \frac{\delta\phi_d}{\lambda_0/D} = \tau\frac{\delta\Omega}{2\pi}, \tag{6.3-2}$$

where $\tau = D/V_s$ is the *transit time* of the sound through the light beam. With the previously calculated $\delta\phi_d \sim 3$ mrad and using a light beam of width $D = 5$ mm, the number of the achievable resolution is ~ 25 spots. For an acoustic transducer of $L = 5$ cm, the Q in this case is about 19, which is well within the Bragg regime. Note that from Eq. (6.3-2) one can achieve improvement in resolution by expanding the lateral width of the light beam traversing the Bragg cell. Because the relation between the deflection angle and the frequency sweep is linear, a simple mechanism for high-speed laser beam scanning may be made possible through the acoustooptic effect because no moving mechanical parts are involved with this kind of scanning mechanisn.

Instead of a single frequency input, the sound cell can be addressed simultaneously by a spectrum of frequencies. The Bragg

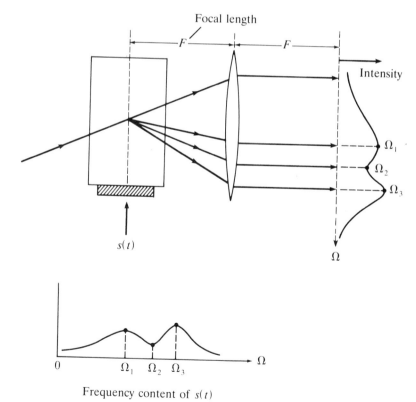

Figure 6.11 An acoustooptic spectrum analyzer.

Acoustooptics and Electrooptics

cell scatters light beams into angles controlled by the spectrum of acoustic frequencies as each frequency generates a beam at a specific diffracted angle. Because the acoustic spectrum is identical to the frequency spectrum of the electrical signal being fed to the cell, the device essentially acts as a *spectrum analyzer*. Figure 6.11 depicts a convenient way to display the frequency content of the electrical signal $s(t)$. Note that the frequency resolution of the analyzer is determined by N.

6.3.3 Demodulation of Frequency Modulated (FM) Signals

From the preceding discussion, we recognize the Bragg cell's frequency-selecting capability. Here we discuss how to make use of this to demodulate FM signals [Pieper and Poon (1985)]. As seen from Figure 6.12, the Bragg cell diffracts light into angles ϕ_{di} controlled by the spectrum of carrier frequencies Ω_{0i}, $i = 1, 2, \ldots$, where each carrier has been frequency-modulated. For the ith FM station, the instantaneous frequency of the signal is representable as $\Omega_i(t) = \Omega_{0i} + \Delta\Omega_i(t)$, where Ω_{0i} is a fixed carrier frequency and $\Delta\Omega_i(t)$ represents a time-varying frequency difference proportional to the amplitude of the modulating signal. As a usual

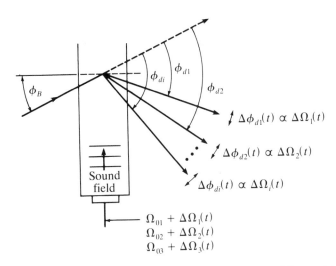

Figure 6.12 Illustration of the nominal directions of the diffracted beams as well as the wobbling due to the frequency modulation.

practice, the FM variation $\Delta\Omega_i(t)$ is small compared to the carrier Ω_{0i}. Using Eq. (6.3-1), the ith FM station is beamed, on the average, in a direction *relative* to the incident beam given by

$$\phi_{di} = \frac{\lambda_0 \Omega_{0i}}{2\pi V_s}. \tag{6.3-3}$$

This is illustrated in Figure 6.12. For each FM carrier, there will now be an independently scattered light beam in a direction determined by the carrier frequency. For clarity, only a few of the scattered light beams are shown. The principle behind the *FM demodulation* is that the actual instantaneous angle of deflection deviates slightly from Eq. (6.3-3) due to the inclusion of $\Delta\Omega_i(t)$, which causes a "wobble" $\Delta\phi_{di}(t)$ in the deflected beam. In particular, we find, with Eq. (6.3-1),

$$\Delta\phi_{di}(t) = \left(\frac{\lambda_0}{2\pi V_s}\right) \Delta\Omega_i(t). \tag{6.3-4}$$

Because, in FM, the frequency variation $\Delta\Omega_i(t)$ is proportional to the amplitude of the audio signal, the variation in the deflected angle $\Delta\phi_{di}(t)$ is likewise proportional to the modulating signal. By placing a knife-edge screen in front of a photodiode positioned along the direction of ϕ_{di}, the amount of light reaching the photodiode, to first order, varies linearly with the small wobble $\Delta\Omega_i(t)$ and hence gives a current proportional to $\Delta\Omega_i(t)$ [see Eq. (6.3-4)], that is, proportional to the amplitude of the modulating signal. In fact, by placing an array of knife-edge–screened detectors, we can monitor all the FM stations simultaneously. This knife-edge technique has been used previously for surface acoustic wave detection [see, for example, Whitman and Korpel (1969)].

6.3.4 Bistable Switching

Bistability refers to the existence of two stable states of a system for a given set of input conditions. Bistable optical devices have received much attention in recent years because of their potential application in optical signal processing. In general, nonlinearity and feedback are required to achieve bistability. Figure 6.13 shows an *acoustooptic bistable device* operating in the Bragg regime.

Acoustooptics and Electrooptics

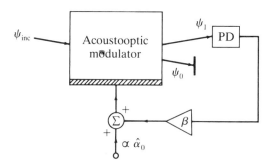

Figure 6.13 Acoustooptic bistable device. PD is a photodetector.

The light diffracted into the first order is detected by the photodetector, amplified, summed with a bias $\hat{\alpha}_0$, and fed back to the acoustic transducer to change the amplitude of its drive signal, which in turn amplitude-modulates the intensities of the diffracted light. Hence the feedback signal has a recursive influence on the diffracted light intensities. Note that the nonlinearity involving only two diffracted orders is a sine-squared function [see Eqs. (6.2-42)]:

$$I_1 = |\psi_1|^2 = I_{inc} \sin^2(\hat{\alpha}/2), \qquad (6.3\text{-}5)$$

where $I_{inc} = |\psi_{inc}|^2$ is the incident intensity and we have a system with a nonlinear input $(\hat{\alpha})$–output (I_1) relationship.

The effective $\hat{\alpha}$ scattering the light in the acoustooptic cell is given by the *feedback equation,*

$$\hat{\alpha} = \hat{\alpha}_0 + \beta I_1, \qquad (6.3\text{-}6)$$

where β denotes the product of the gain constant of the amplifier and the quantum efficiency of the photodetector (PD) (see Section 7.6).

Note that under the feedback action, $\hat{\alpha}$ can no longer, in general, be treated as a constant. In fact, $\hat{\alpha}$ can be treated as constant during interaction if the interaction time, given as the ratio of the laser beam width and the bulk speed of sound in the cell, is very small compared to the delays incorporated by the finite response time of the photodetector (see Section 7.6), the sound-cell driver and the feedback amplifier, or any other delay line that may

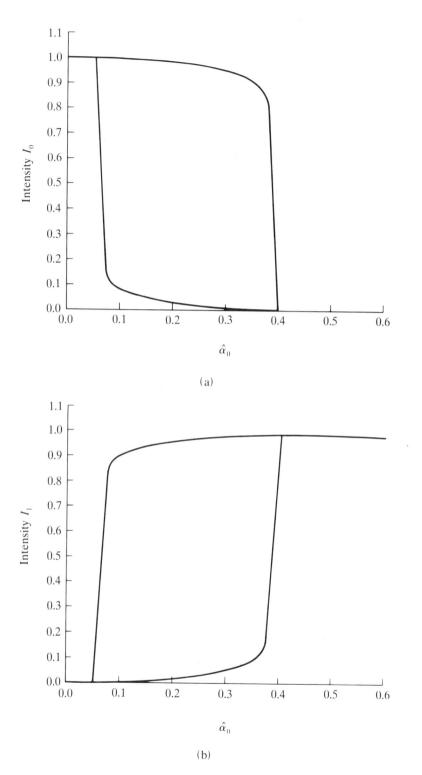

Figure 6.14 Hysteresis curve (intensity vs. bias $\hat{\alpha}_0$), $\beta = 2.6$ and $Q = 20$: (a) for the zeroth order light, I_0; (b) for the first-order light I_1.

be purposely installed (e.g., an optical fiber or coaxial cable) in the feedback path. We consider this case only. The steady-state behavior of the system is then given by the simultaneous solution of Eqs. (6.3-5) and (6.3-6) and is an S-shaped curve if we would plot I_1 vs. $\hat{\alpha}_0$ exactly. The stability of different parts of similar curves will be discussed in Section 8.4. The hysteretic behavior can be numerically found by programming (6.3-5) with (6.3-6) in a computer and calculating the steady state values of I_1 (or $I_0 = |\psi_0|^2$) for every small increment in $\hat{\alpha}_0$. This is shown in Figure 6.14 [Banerjee and Poon (1987), Poon and Cheung (1989)]. Note that, from Figure 6.14(b), a gradual increase in input bias $\hat{\alpha}_0$ produces a steady increase in the output intensity I_1 (which represents the *lower stable state*) until reaching a critical value where the output switches up to the *higher stable state*. On decreasing the input, the output does not immediately fall sharply but remains on the upper branch of the curve (the higher stable state) until the input is reduced to a lower critical value, at which the output switches down. The difference in the values of $\hat{\alpha}_0$ at which the transitions occur gives rise to the so-called *hysteresis*. In passing, we also mention that similar hysteretic behavior may be observed by treating I_{inc} or β as the input and treating the other variables such as $\hat{\alpha}_0$ as parameters. We discuss another type of bistable system exhibiting hysteresis in Section 8.4.

6.4 Wave Propagation in Anisotropic Media

Thus far, in this book, we have studied the effects of wave propagation through isotropic media. However, many materials (e.g., crystals) are anisotropic. In this section, we will study linear wave propagation in a medium that is homogeneous and magnetically isotropic (μ_0 constant) but that allows for electrical anisotropy. By this we mean that the polarization produced in the medium by an applied electric field is no longer just a constant times the field, but depends critically on the direction of the applied field in relation to the anisotropy of the medium. This discussion will help us understand the properties and uses of electrooptic materials, to be introduced in Sections 6.5 and 6.6.

6.4.1 The Dielectric Tensor

Figure 6.15 depicts a model illustrating anisotropic binding of an electron in a crystal. *Anisotropy* is taken into account by

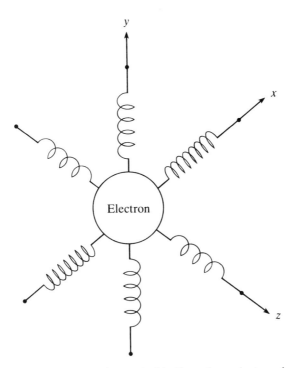

Figure 6.15 Model illustrating anisotropic binding of an electron in a crystal (the spring constants are different in each direction).

assuming different spring constants in each direction. (In the isotropic case, all spring constants are equal). Consequently, the displacement of the electron under the influence of an external electric field depends not only on the magnitude of the field but also on its direction. It follows, in general, that the vector **D** will no longer be the direction of **E**. Thus, in place of Eq. (3.1-13a), the components of **D** and **E** are related by the following equation [Eq. (3.1-13b) still holds true, as we are considering a magnetically isotropic medium]:

$$D_x = \epsilon_{xx} E_x + \epsilon_{xy} E_y + \epsilon_{xz} E_z, \qquad (6.4\text{-}1a)$$

$$D_y = \epsilon_{yx} E_x + \epsilon_{yy} E_y + \epsilon_{yz} E_z, \qquad (6.4\text{-}1b)$$

$$D_z = \epsilon_{zx} E_x + \epsilon_{zy} E_y + \epsilon_{zz} E_z, \qquad (6.4\text{-}1c)$$

or

$$D_i = \sum_{j=1}^{3} \epsilon_{ij} E_j, \qquad (6.4\text{-}2)$$

where $i, j = 1$ for x, 2 for y, and 3 for z. Equations (6.4-1a)–(6.4-1c), or Eq. (6.4-2), are customarily written as

$$\mathbf{D} = \epsilon \mathbf{E}, \qquad (6.4\text{-}3)$$

where $\mathbf{D} = [D_1 \; D_2 \; D_3]^T$, $\mathbf{E} = [E_1 \; E_2 \; E_3]^T$, and

$$\epsilon = \begin{bmatrix} \epsilon_{11} & \varepsilon_{12} & \epsilon_{13} \\ \epsilon_{21} & \epsilon_{22} & \epsilon_{23} \\ \epsilon_{31} & \epsilon_{32} & \epsilon_{33} \end{bmatrix}, \qquad (6.4\text{-}4)$$

or simply as

$$D_i = \epsilon_{ij} E_j. \qquad (6.4\text{-}5)$$

The 3×3 matrix in Eq. (6.4-4) is commonly known as the *dielectric tensor*. Equation (6.4-5) is merely a shorthand representation of Eq. (6.4-2), using the Einstein convention (see Chapter 1). We remind ourselves that the Einstein convention assumes an implied summation over repeated indices (viz., j) on the same side (RHS or LHS) of an equation.

In a lossless medium, the dielectric tensor is symmetric, that is,

$$\epsilon_{ij} = \epsilon_{ji}, \qquad (6.4\text{-}6)$$

and has only six independent elements [see, for instance, Haus (1984) for proof].

Now, it is well known that any real symmetric matrix can be diagonalized through a coordinate transformation. Hence, the dielectric tensor can assume the diagonal form

$$\epsilon = \begin{bmatrix} \epsilon_x & 0 & 0 \\ 0 & \epsilon_y & 0 \\ 0 & 0 & \epsilon_z \end{bmatrix}. \qquad (6.4\text{-}7)$$

Table 6.1 Crystal Classes and Some Common Examples

	Cubic	Uniaxial	Biaxial
Principal axis system	$\begin{bmatrix} \epsilon & 0 & 0 \\ 0 & \epsilon & 0 \\ 0 & 0 & \epsilon \end{bmatrix}$	$\begin{bmatrix} \epsilon_x & 0 & 0 \\ 0 & \epsilon_x & 0 \\ 0 & 0 & \epsilon_z \end{bmatrix}$	$\begin{bmatrix} \epsilon_x & 0 & 0 \\ 0 & \epsilon_y & 0 \\ 0 & 0 & \epsilon_z \end{bmatrix}$
Common examples	Sodium chloride Diamond	Quartz (positive, $\epsilon_x < \epsilon_z$) Calcite (negative, $\epsilon_x > \epsilon_z$)	Mica Topaz

The new coordinate system is called the *principal axis system*, the three ϵs are known as the *principal dielectric constants*, and the Cartesian coordinate axes are called the *principal axes*. Three crystal classes (shown in Table 6.1) can be identified in terms of Eq. (6.4-7): *cubic*, *uniaxial*, and *biaxial*. Because most of the crystals used for electrooptic devices are uniaxial, we will therefore concentrate only on these types of crystals in our subsequent discussions. Note that for uniaxial crystals, the axis, which is characterized by a component of $\epsilon = \epsilon_z$, is called the *optic axis*. When $\epsilon_z > \epsilon_x = \epsilon_y$, the crystal is *positive uniaxial*, and when $\epsilon_z < \epsilon_x = \epsilon_y$, it is *negative uniaxial*. A word on notation: We will use subscripts x, y, z instead of $1, 2, 3$ in all following discussions unless otherwise stated.

6.4.2 Plane-Wave Propagation in Uniaxial Crystals; Birefringence

In order to advance a consolidated treatment of plane-wave propagation in uniaxial crystals, it is advantageous to first describe, in general, the expression for **D** in terms of **E**. This can be achieved by rewriting Eqs. (3.1-3) and (3.1-4) and realizing that the operators $\partial/\partial t$ and ∇ may be replaced according to

$$\frac{\partial}{\partial t} \rightarrow j\omega_0$$

$$\nabla \rightarrow -jk_0 \mathbf{a}_k = -j\mathbf{k}_0 \tag{6.4-8}$$

if and only if all the dependent variables in Maxwell's equations, namely, **H**, **B**, **D**, and **E**, vary according to $\exp[j(\omega_0 t - \mathbf{k}_0 \cdot \mathbf{r})]$ and have constant amplitudes. Using Eqs. (6.4-8) in Eqs. (3.1-3) and

(3.1-4), and assuming $\mathbf{B} = \mu_0 \mathbf{H}$, we get

$$\mathbf{H} = \frac{k_0 \mathbf{a}_k \times \mathbf{E}}{\omega_0 \mu_0} \qquad (6.4\text{-}9)$$

and

$$\mathbf{D} = -\frac{k_0 \mathbf{a}_k \times \mathbf{H}}{\omega_0}, \qquad (6.4\text{-}10)$$

where $\mathbf{a}_k = \mathbf{k}_0 / |\mathbf{k}_0| = \mathbf{k}_0 / k_0$. Finally, eliminating $\mathbf{H}$ between Eqs. (6.4-9) and (6.4-10), we obtain

$$\mathbf{D} = \frac{k_0^2}{\omega_0^2 \mu_0} \left[\mathbf{E} - (\mathbf{a}_k \cdot \mathbf{E}) \mathbf{a}_k \right], \qquad (6.4\text{-}11)$$

where we have used the vector identity

$$\mathbf{A} \times (\mathbf{C} \times \mathbf{A}) = \mathbf{C} - (\mathbf{A} \cdot \mathbf{C}) \mathbf{A}. \qquad (6.4\text{-}12)$$

For electrically isotropic media, we know that

$$\mathbf{D} = \frac{k_0^2}{\omega_0^2 \mu_0} \mathbf{E} = \frac{1}{v^2 \mu_0} \mathbf{E} = \epsilon \mathbf{E} \qquad (6.4\text{-}13)$$

and hence $\mathbf{a}_k \cdot \mathbf{E} = 0$, that is, there is no component of the electric field along the direction of propagation (see Chapter 3). However, in anisotropic crystals, $\mathbf{D} = \epsilon \mathbf{E}$, [see Eq. (6.4-3)]; hence Eq. (6.4-11) becomes

$$\epsilon \mathbf{E} = \frac{k_0^2}{\omega_0^2 \mu_0} \left[\mathbf{E} - (\mathbf{a}_k \cdot \mathbf{E}) \mathbf{a}_k \right]. \qquad (6.4\text{-}14)$$

Incidentally, from the first of Maxwell's equations [Eq. (3.1-1)], it follows, using Eq. (6.4-8), that $\mathbf{a}_k \cdot \mathbf{D} = 0$. This means that for a general anisotropic medium, $\mathbf{D}$ is perpendicular to the direction of propagation, even though $\mathbf{E}$ may *not* be so because of the anisotropy. This is summarized in Figure 6.16. Note that the Poynting vector $\mathbf{S} = \mathbf{E} \times \mathbf{H}$ (which determines the direction of

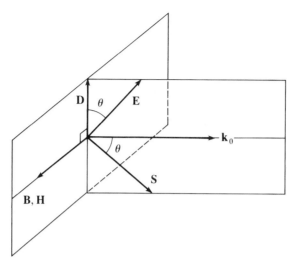

Figure 6.16 Diagram illustrating the direction of various field quantities in anisotropic crystals ($S = E \times H$ is the Poynting vector).

energy flow) is different from the direction of propagation of the wavefronts denoted by $\mathbf{k}_0$.

We will now use Eq. (6.4-14) to analyze plane-wave propagation in uniaxial crystals. Suppose that a plane wave is polarized so that its electric field $\mathbf{E}_a$ is perpendicular to the optic axis, and the propagation vector is at an angle θ with respect to the optic axis as shown in Figure 6.17. The y direction is pointing normally outward from the paper, hence $\mathbf{E}_a = E_a \mathbf{a}_y$. Substituting into Eq. (6.4-14), we can find the propagation constant $k_0^2 = \omega_0^2 \mu_0 \epsilon_y = \omega_0^2 \mu_0 \epsilon_x$ and hence the phase velocity $v_2 = 1/\sqrt{\mu_0 \epsilon_y} = 1/\sqrt{\mu_0 \epsilon_x} = v_1$, corresponding to the wave polarized along the y direction. It can be shown that for any $\mathbf{E}$ that has no component along the optic axis, the same velocity (v_1) is obtained.

Consider a second example, where $\mathbf{E}_b = E_{bx} \mathbf{a}_x + E_{bz} \mathbf{a}_z$ and where the direction of $\mathbf{E}_b$ is normal to the direction of propagation $\mathbf{a}_k$. Equation (6.4-14) then reduces to the two component equations

$$\omega_0^2 \mu_0 \epsilon_x E_{bx} = k_0^2 \left[E_{bx} - (\sin \theta E_{bx} + \cos \theta E_{bz}) \sin \theta \right], \quad (6.4\text{-}15a)$$

$$\omega_0^2 \mu_0 \epsilon_z E_{bz} = k_0^2 \left[E_{bz} - (\sin \theta E_{bx} + \cos \theta E_{bz}) \cos \theta \right] \quad (6.4\text{-}15b)$$

Acoustooptics and Electrooptics

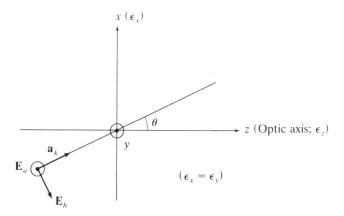

Figure 6.17 Plane-wave propagation in a uniaxial crystal.

where a_k in Eq. (6.4-14) has been written as $\mathbf{a}_k = \sin\theta\,\mathbf{a}_x + \cos\theta\,\mathbf{a}_z$. We can reexpress Eqs. (6.4-15a) and (6.4-15b) as

$$
\begin{bmatrix} \omega_0^2\mu_0\epsilon_x - k_0^2\cos\theta & k_0^2\cos\theta\sin\theta \\ k_0^2\cos\theta\sin\theta & \omega_0^2\mu_0\epsilon_z - k_0^2\sin^2\theta \end{bmatrix}\begin{bmatrix} E_{bx} \\ E_{bz} \end{bmatrix} = 0. \quad (6.4\text{-}16)
$$

For nontrivial solutions, the determinant of this matrix should be zero, and this gives

$$
k_0^2 = \frac{k_{0x}^2 k_{0z}^2}{k_{0x}^2\sin^2\theta + k_{0z}^2\cos^2\theta}, \quad (6.4\text{-}17)
$$

where $k_{0x}^2 = \omega_0^2\mu_0\epsilon_x$ and $k_{0z}^2 = \omega_0^2\mu_0\epsilon_z$. The total phase velocity is accordingly given by

$$
v = \sqrt{(v_1\cos\theta)^2 + (v_3\sin\theta)^2}, \quad (6.4\text{-}18)
$$

where $v_1 = 1/\sqrt{\mu_0\epsilon_x}$ and $v_3 = 1/\sqrt{\mu_0\epsilon_z}$. Note that when $\theta = 0$, the wave propagates along the optic axis with the electric field (in the $-\mathbf{a}_x$ direction) perpendicular to it. The phase velocity is $v_1 = 1/\sqrt{\mu_0\epsilon_x}$. In fact, the phase velocity of the wave is the same for any wave that has no polarization vector component along the optic axis, that is, $v_1 = v_2 = 1/\sqrt{\mu_0 \epsilon_x} = c/n_o$ as $\epsilon_x = \epsilon_y$ in uniaxial crystals, where $n_o = \sqrt{\epsilon_x/\epsilon_0}$ is the *ordinary refractive index*. The wave associated with the two identical refractive indices is often called the *ordinary wave*. Now, when $\theta = \pi/2$, the polarization vector lies along the optic axis. The phase velocity of the wave is $v_3 = 1/\sqrt{\mu_0 \epsilon_z} = c/n_e$, where $n_e = \sqrt{\epsilon_z/\epsilon_0}$ is the *extraordinary refractive index* and the wave associated with the dissimilar refractive index is the *extraordinary wave*. This phenomenon, in which the phase velocity of an optical wave propagating in the crystal depends on the direction of its polarization, is called *birefringence*. It can be shown that when a light beam is incident on a uniaxial crystal in a direction other than the optic axis, two beams will emerge from the crystal, due to the two dissimilar indices of refraction (n_e, n_o), and both are polarized perpendicular to each other. Therefore, if we view an object through the crystal, two images will be observed. This remarkable optical phenomenon in crystals is called *double refraction*.

6.4.3 Applications of Birefringence: Wave Plates

Consider a plate made of a uniaxial material. The optic axis is along the z direction, as shown in Figure 6.18. Let a linearly polarized real optical field incident on the crystal at $x = 0$ cause a field in the crystal at $x = 0^+$ of the form

$$\mathbf{E}_{\text{inc}} = \text{Re}\left[E_0(\mathbf{a}_y + \mathbf{a}_z)e^{j\omega_0 t} \right]. \tag{6.4-19}$$

The wave travels through the plate, and at the right edge of the plate ($x = d$) the real field can be represented as

$$\mathbf{E}_{\text{out}} = \text{Re}\left[E_0\left(\exp\left(-j\frac{\omega_0}{v_2}d \right)\mathbf{a}_y + \exp\left(-j\frac{\omega_0}{v_3}d \right)\mathbf{a}_z \right)\exp(j\omega_0 t) \right]. \tag{6.4-20}$$

Note that the two plane-polarized waves acquire a different phase as they propagate through the crystal. The relative phase shift $\Delta\phi$

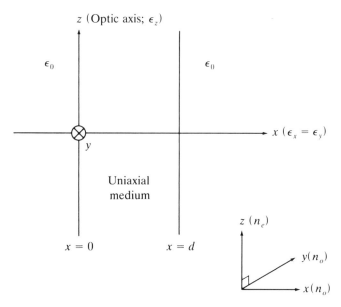

Figure 6.18 A wave plate (retardation plate).

between the extraordinary and ordinary waves is

$$\Delta\phi = \left(\frac{\omega_0}{v_3} - \frac{\omega_0}{v_2}\right)d$$

$$= \frac{\omega_0}{c}(n_e - n_o)d$$

$$= \frac{2\pi}{\lambda_v}(n_e - n_o)d, \qquad (6.4\text{-}21)$$

where λ_v is the wavelength of light in vacuum. If $n_e > n_o$, the extraordinary wave lags the ordinary wave in phase, that is, the ordinary wave travels faster, whereas if $n_e < n_o$, the opposite is true. Such a phase shifter is often referred to as a *compensator* or a *retardation plate*. The directions of polarization for the two allowed waves are mutually orthogonal and are usually called the *slow* and *fast axes* of the crystal. If $n_e > n_o$, the z axis is the slow axis and the y axis is the fast axis. Since the plate surfaces are perpendicular to

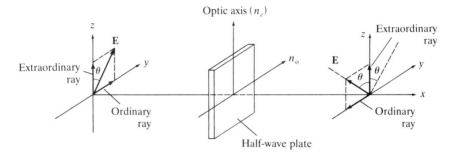

Figure 6.19 Rotation of polarization by a half-wave plate.

the principal x axis, the direction in which the wave propagates, the crystal plate is then called *x-cut*. If $\Delta\phi = 2m\pi$, where m is an integer (by introducing the right thickness d), the compensator is called a *full-wave plate*. The extraordinary wave is retarded or advanced by a full cycle of λ_v with respect to the ordinary wave. The emerging beam will have the same state of polarization as the incident beam. If $\Delta\phi = (2m + 1)\pi$ and $(m + \frac{1}{2})\pi$, we have a *half-wave plate* and a *quarter-wave plate*, respectively. A half-wave plate can rotate the plane of polarization of the incident wave. This is illustrated in Figure 6.19. Note that a relative phase change of π is equivalent to reversing one of the components of **E**. The combination of the two components at the exit of the plate shows that the polarization direction has been rotated through twice the angle θ. It should also be noted that the rotation is independent of which component is reversed, as the polarization direction only specifies the line along which the vector **E** oscillates. Although a half-wave plate can rotate the polarization direction, a quarter-wave plate can transform a plane-polarized wave to a circularly polarized wave, as the plate introduces a relative phase shift of $\pi/2$ between the extraordinary and ordinary waves. Finally, it should be noted that for any other value of $\Delta\phi$, the emerging wave will be elliptically polarized for a plane-polarized incident wave.

6.4.4 The Index Ellipsoid

We have shown mathematically that for a plane-polarized wave propagating in any given direction in a uniaxial crystal, there are *two allowed polarizations*, one along the optic axis and the other perpendicular to it. As shown in Section 6.4.2, the total phase velocity of a wave propagating in an arbitrary direction depends on the velocities of waves polarized solely along the directions of the principal axes and on the direction of propagation of the wave. A convenient method to figure out the directions of polarization of the two allowed waves and their phase velocities is through the *index ellipsoid*, a mathematical entity written as

$$\frac{x^2}{n_x^2} + \frac{y^2}{n_y^2} + \frac{z^2}{n_z^2} = 1, \tag{6.4-22}$$

where $n_x^2 = \epsilon_x/\epsilon_0$, $n_y^2 = \epsilon_y/\epsilon_0$, and $n_z^2 = \epsilon_z/\epsilon_0$. Figure 6.20(a) shows the index ellipsoid. From Eq. (6.4-22), we can determine the respective refractive indices as well as the two allowed polarizations of **D** for a given direction of propagation in crystals. To see how this can be done, consider a plane perpendicular to the direction of the propagation $\mathbf{k}_0$, containing the center of the ellipsoid. The intersection of the plane with the ellipsoid is the ellipse A in Figure 6.20(b), drawn for a uniaxial crystal. The directions of the two possible displacements $\mathbf{D}_1$ and $\mathbf{D}_2$ now coincide with the major and minor axes of the ellipse A, and the appropriate refractive indices

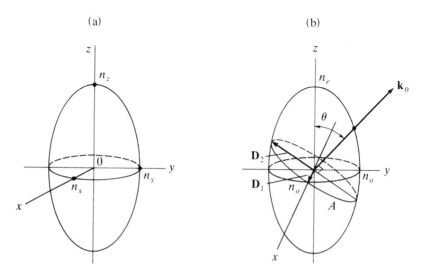

Figure 6.20 (a) Index ellipsoid. (b) Two allowed polarizations $\mathbf{D}_1$ and $\mathbf{D}_2$.

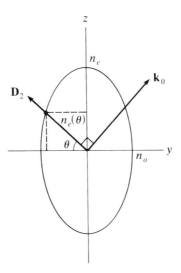

Figure 6.21 Cross section of index ellipsoid illustrating the value of refractive index depending on the direction of wave propagation.

for the two allowed plane-polarized waves are given by the lengths of the two semiaxes. For uniaxial crystals $\epsilon_x = \epsilon_y$, and ϵ_z is distinct. We then have $n_x = n_y \equiv n_o$ and $n_z \equiv n_e$. $\mathbf{D}_1$ is the ordinary wave and $\mathbf{D}_2$ is the extraordinary wave. The index of refraction $n_e(\theta)$ along $\mathbf{D}_2$ can be determined using Figure 6.21. Employing the relations

$$n_e^2(\theta) = z^2 + y^2 \qquad (6.4\text{-}23a)$$

$$\frac{z}{n_e(\theta)} = \sin \theta, \qquad (6.4\text{-}23b)$$

and the equation of the ellipse

$$\frac{y^2}{n_o^2} + \frac{z^2}{n_e^2} = 1, \qquad (6.4\text{-}23c)$$

we have

$$\frac{1}{n_e^2(\theta)} = \frac{\cos^2 \theta}{n_o^2} + \frac{\sin^2 \theta}{n_e^2}. \qquad (6.4\text{-}24)$$

Note that Eq. (6.4-24) can be obtained directly from Eq. (6.4-18). From Figure 6.21 we can observe immediately that when $\theta = 0$, that is, the wave is propagating along the optic axis, no birefringence is observed $[n_e(0) - n_o = 0]$. Also, the amount of birefringence, $n_e(\theta) - n_o$, depends on the propagation direction and it is maximum $(= n_e - n_o)$ when the propagation direction is perpendicular to the optic axis, $\theta = 90°$.

6.5 Electrooptic Effect in Uniaxial Crystals

Having been introduced to wave propagation in anisotropic media, we are now in a position to analyze the electrooptic effect, which is inherently anisotropic. As seen later, we can effectively study this using the index elliposoid concepts. In the following section, we will study the applications of the electrooptic effect in areas discussed in connection with acoustooptics, for example, amplitude and phase modulation.

The *electrooptic effect* is, loosely speaking, the change in n_e and n_o that is caused by an applied electric field. In this chapter, we will only discuss the *linear* (or *Pockels-type*) *electrooptic effect*, meaning the case where the change in n_e and n_o is linearly proportional to the applied field. Note that the *Pockels effect* can only exist in some crystals, namely, those that do not possess inversion symmetry [see Yariv (1976)]. The other case, namely, where n_e and n_o depend nonlinearly on the applied field, is called the *Kerr effect* and will be discussed in the context of nonlinear optics in Chapter 8.

Mathematically, the electrooptic effect can be best represented as a deformation of the index ellipsoid due to an external electric field. Thus, Eq. (6.4-22), with $n_x = n_y = n_o$ and $n_z = n_e$, represents the ellipsoid for uniaxial crystals in the absence of an applied field, that is,

$$\frac{x^2}{n_o^2} + \frac{y^2}{n_o^2} + \frac{z^2}{n_e^2} = 1, \qquad (6.5\text{-}1)$$

where the directions x, y, and z are the principal axes.

Restricting our analysis to the linear electrooptic (Pockels) effects, the general expression for the deformed ellipsoid is

$$\left[\frac{1}{n_o^2} + \Delta\left(\frac{1}{n^2}\right)_1\right]x^2 + \left[\frac{1}{n_o^2} + \Delta\left(\frac{1}{n^2}\right)_2\right]y^2 + \left[\frac{1}{n_e^2} + \Delta\left(\frac{1}{n^2}\right)_3\right]z^2$$

$$+ 2\,\Delta\left(\frac{1}{n^2}\right)_4 yz + 2\,\Delta\left(\frac{1}{n^2}\right)_5 xz + 2\,\Delta\left(\frac{1}{n^2}\right)_6 xy = 1, \quad (6.5\text{-}2)$$

where

$$\Delta\left(\frac{1}{n^2}\right)_i = \sum_{j=1}^{3} r_{ij}\mathscr{E}_j, \qquad i = 1,\ldots,6, \quad (6.5\text{-}3)$$

and where r_{ij} are called linear electrooptic (or Pockels) coefficients. The $\mathscr{E}_j$s are the components of the externally applied electric field in the x, y, and z directions (to be distinguished from the optical fields represented as E_js). We can express Eq. (6.5-3) in matrix form as

$$\begin{bmatrix} \Delta\left(\dfrac{1}{n^2}\right)_1 \\[2mm] \Delta\left(\dfrac{1}{n^2}\right)_2 \\[2mm] \Delta\left(\dfrac{1}{n^2}\right)_3 \\[2mm] \Delta\left(\dfrac{1}{n^2}\right)_4 \\[2mm] \Delta\left(\dfrac{1}{n^2}\right)_5 \\[2mm] \Delta\left(\dfrac{1}{n^2}\right)_6 \end{bmatrix} = \begin{bmatrix} r_{11} & r_{12} & r_{13} \\ r_{21} & r_{22} & r_{23} \\ r_{31} & r_{32} & r_{33} \\ r_{41} & r_{42} & r_{43} \\ r_{51} & r_{52} & r_{53} \\ r_{61} & r_{62} & r_{63} \end{bmatrix} \begin{bmatrix} \mathscr{E}_1 \\ \mathscr{E}_2 \\ \mathscr{E}_3 \end{bmatrix}. \quad (6.5\text{-}4)$$

In Eqs. (6.5-2)–(6.5-4), we have tacitly gone to the *other* convention (viz., 1, 2, 3 instead of x, y, z) to comply with standard nomenclature. Note that when the applied field is zero, Eq. (6.5-2) reduces to

Acoustooptics and Electrooptics

Table 6.2 Electrooptic Coefficients

Material	r_{ij} (10^{-12} m / V)	λ_v (μm)	Refractive Index
LiNbO$_3$	$r_{13} = r_{23} = 8.6$	0.63	$n_o = 2.286$
	$r_{33} = 30.8$		$n_e = 2.200$
	$r_{22} = -r_{61} = -r_{12} = 3.4$		
	$r_{51} = r_{42} = 28$		
SiO$_2$	$r_{11} = -r_{21} = -r_{62} = 0.29$	0.63	$n_o = 1.546$
	$r_{41} = -r_{52} = 0.2$		$n_e = 1.555$
KDP	$r_{41} = r_{52} = 8.6$	0.55	$n_o = 1.51$
(Potassium	$r_{63} = 10.6$		$n_e = 1.47$
dihydrogen			
phosphate)			
ADP	$r_{41} = r_{52} = 2.8$	0.55	$n_o = 1.52$
(Ammonium	$r_{63} = 8.5$	0.55	$n_e = 1.48$
dihydrogen			
phosphate)			
GaAs	$r_{41} = r_{52} = r_{63} = 1.2$	0.9	$n_o = n_e = 3.42$
			(Cubic)

Eq. (6.5-1). Equation (6.5-4) contains 18 elements and they are necessary in the most general case, when no symmetry is present in the crystal. Otherwise, some of the elements are zero and/or some of the nonzero elements have the same value. Table 6.2 lists the nonzero elements of the linear electrooptic coefficients of some commonly used crystals. Using Eqs. (6.5-2) and (6.5-4) and Table 6.2, we can find the equation of the index ellipsoid in the presence of an external applied field. For instance, for a KDP crystal in the presence of an external field $\mathscr{E} = \mathscr{E}_x \mathbf{a}_x + \mathscr{E}_y \mathbf{a}_y + \mathscr{E}_z \mathbf{a}_z$, the index ellipsoid equation can be reduced to

$$\frac{x^2}{n_o^2} + \frac{y^2}{n_o^2} + \frac{z^2}{n_e^2} + 2r_{41}\mathscr{E}_x yz + 2r_{41}\mathscr{E}_y xz + 2r_{63}\mathscr{E}_z xy = 1. \quad (6.5\text{-}5)$$

The mixed terms in the equation of the index ellipsoid imply that the major and minor axes of the ellipsoid, with a field applied, are no longer parallel to the x, y, and z axes, which are the directions of the principal axes when no field is present. This deformation of the index ellipsoid creates the externally induced birefringence.

6.6 Some Applications of the Electrooptic Effect

6.6.1 Amplitude Modulation

Longitudinal Configuration

A typical arrangement of an electrooptic *amplitude modula-tor* is shown in Figure 6.22. It consists of a electrooptic crystal placed between two crossed polarizers whose polarization axes are perpendicular to each other. The *polarization axis* defines the direction along which emerging light is linearly polarized. We use a KDP crystal with its principal axes aligned with x, y, and z. An electric field is applied through the voltage V along the z-axis, which is the direction of propagation of the optical field, thus justifying the name *longitudinal configuration*. Eq. (6.5-5) then becomes

$$\frac{x^2}{n_o^2} + \frac{y^2}{n_o^2} + \frac{z^2}{n_e^2} + 2r_{63}\mathscr{E}_z xy = 1. \tag{6.6-1}$$

By inspection, the ellipsoid has principal axes along the coordinates x', y', and z, where

$$x' = \frac{1}{\sqrt{2}}(x - y), \qquad y' = \frac{1}{\sqrt{2}}(x + y). \tag{6.6-2}$$

Introducing these expressions into Eq. (6.6-1), we have the index

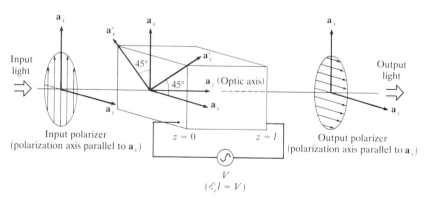

Figure 6.22 A longitudinal electrooptic intensity modulation system.

ellipsoid equation in the coordinate system aligned with the new principal axes:

$$\left(\frac{1}{n'_x}\right)^2 x'^2 + \left(\frac{1}{n'_y}\right)^2 y'^2 + \frac{z^2}{n_e^2} = 1. \qquad (6.6\text{-}3)$$

In Eq. (6.6-3),

$$\frac{1}{n'^2_x} = \frac{1}{n^2_o} - r_{63}\mathcal{E}_z, \qquad (6.6\text{-}4)$$

implying

$$n'_x \approx n_o + \frac{n_o^3}{2}r_{63}\mathcal{E}_z, \qquad (6.6\text{-}5)$$

assuming $n_o^{-2} \gg r_{63}\mathcal{E}_z$. Similarly,

$$n'_y \approx n_o - \frac{n_o^3}{2}r_{63}\mathcal{E}_z. \qquad (6.6\text{-}6)$$

Also,

$$n_z = n_e. \qquad (6.6\text{-}7)$$

The input (optical) field $\text{Re}\{E_0 e^{j\omega_0 t}\}\mathbf{a}_x$ at $z = 0$ is polarized along the x direction and can be resolved into two mutually orthogonal components polarized along x' and y'. After passage through the electrooptic crystal, the field components are

$$E'_x|_{z=l} = \text{Re}\left[\frac{E_0}{\sqrt{2}}\exp\left(j\left[\omega_0 t - \frac{\omega_0}{c}n'_x l\right]\right)\right], \qquad (6.6\text{-}8a)$$

$$E'_y|_{z=l} = \text{Re}\left[\frac{E_0}{\sqrt{2}}\exp\left(j\left[\omega_0 t - \frac{\omega_0}{c}n'_y l\right]\right)\right]. \qquad (6.6\text{-}8b)$$

The phase difference at $z = l$ between the two components is called the *retardation* Φ_L and is given by

$$\Phi_L = \frac{\omega_0}{c}(n'_x - n'_y)l$$

$$= \frac{\omega_0}{c}n_o^3 r_{63}V = \frac{2\pi}{\lambda_v}n_o^3 r_{63}V, \qquad (6.6\text{-}9)$$

where $V = \mathscr{E}_z l$. At this point it is interesting to mention that at $\Phi_L = \pi$, the electrooptic crystal essentially acts as a half-wave plate where the birefringence is induced electrically. The crystal causes a x-polarized wave at $z = 0$ to acquire a y-polarization at $z = l$. The input light field then passes through the output polarizer unattenuated. With the electric field inside the crystal turned off ($V = 0$), there is no output light, as it is blocked off by the crossed output polarizer. The system, therefore, can switch light on and off electrooptically. The voltage yielding a retardation $\Phi_L = \pi$ is often referred to as the *half-wave voltage*,

$$V_\pi = \frac{\lambda_v}{2n_o^3 r_{63}}. \qquad (6.6\text{-}10)$$

From Table 6.2 and at $\lambda_v = 0.55\ \mu$m, $V_\pi \approx 7.5$ kV for KDP.

Returning to analyze the general system, the **E** component parallel to $\mathbf{a}_y$, that is, the component passed by the output polarizer, is

$$E_y\Big|_{z=l} = \left.\frac{E'_x - E'_y}{\sqrt{2}}\right|_{z=l} = \frac{1}{2}\text{Re}\left\{E_0\left[\exp\left(j\left[\omega_0 t - \frac{\omega_0}{c}n'_x l\right]\right)\right.\right.$$

$$\left.\left. - \exp\left(j\left[\omega_0 t - \frac{\omega_0}{c}n'_y l\right]\right)\right]\right\}. \qquad (6.6\text{-}11)$$

The ratio of the output ($I_0 = E_y^2$) and the input ($I_i = E_0^2$) intensities is

$$\frac{I_0}{I_i} = \sin^2\left[\frac{\omega_0}{2c}(n'_x - n'_y)l\right] = \sin^2\left(\frac{\Phi_L}{2}\right) = \sin^2\left(\frac{\pi}{2}\frac{V}{V_\pi}\right). \qquad (6.6\text{-}12)$$

Figure 6.23 shows a plot of the transmission factor (I_0/I_i) versus

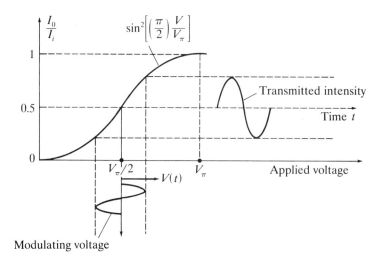

Figure 6.23 Relationship between the modulating signal voltage $V(t)$ and the output intensity.

the applied voltage. Note that the most linear region of the curve is obtained for a bias voltage at $V_\pi/2$. Therefore, the electrooptic modulator is usually biased with a fixed retardation to the 50% transmission point. The bias can be achieved either electrically (by applying a fixed voltage $V = \frac{1}{2}V_\pi$) or optically (by using a quarter-wave plate). The quarter-wave plate has to be inserted between the electrooptic crystal and the output polarizer in such a way that its slow and fast axes are aligned in the direction of $\mathbf{a}'_x$ and $\mathbf{a}'_y$.

Transverse Configuration

In the previous discussion, the electric field was applied along the direction of light propagation. Consider, now, the configuration shown in Figure 6.24, where the applied field is perpendicular to the direction of propagation. In this arrangement, the modulator operates in the *transverse configuration*. The input light propagates along y', with its polarization in the $x'-z$ plane at 45° from the z axis, and the output polarizer is perpendicular to the incident polarization. Due to the applied voltage V, the principal axes x and y of the index ellipsoid are rotated to new principal axes along the x' and y' directions. Again, the coordinate systems are related through Eqs. (6.6-2) and the change of the index

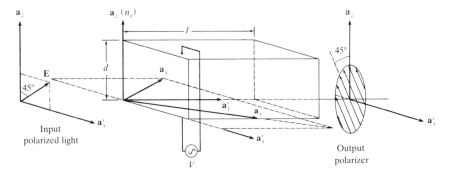

Figure 6.24 A transverse electrooptic intensity modulation system.

ellipsoid produced by $\mathcal{E}_z$ is given by Eq. (6.6-1). Note that, in this configuration, there is no birefringence induced along the z axis. Therefore, the z component of the incident electric field experiences no phase modulation. However, the $\mathbf{a}'_x$ component will experience retardation. Assuming that the input field polarization is $\mathbf{E} = E_0(\mathbf{a}'_x + \mathbf{a}_z)/\sqrt{2}$, the field components at the exit of the crystals are then

$$E'_x = \operatorname{Re}\left\{ \frac{E_0}{\sqrt{2}} \exp\left(j\left[\omega_0 t - \frac{\omega_0}{c} n'_x l \right] \right) \right\} \qquad (6.6\text{-}13)$$

and

$$E_z = \operatorname{Re}\left\{ \frac{E_0}{\sqrt{2}} \exp\left(j\left[\omega_0 t - \frac{\omega_0}{c} n_e l \right] \right) \right\}. \qquad (6.6\text{-}14)$$

The phase difference between the two components is

$$\begin{aligned}
\Phi_T &= \frac{\omega_0}{c}(n'_x - n_e)l \\[2mm]
&= \frac{\omega_0}{c}\left[(n_o - n_e) + \frac{n_o^3}{2} r_{63}\mathcal{E}_z \right] l \quad \text{(using Eq. (6.6-5))} \\[2mm]
&= \frac{\omega_0}{c} l \left[(n_o - n_e) + \frac{n_o^3}{2} r_{63} \frac{V}{d} \right], \qquad (6.6\text{-}15)
\end{aligned}$$

Acoustooptics and Electrooptics

where d is the crystal length along the direction of the applied V. Note that Φ_T can be increased by employing a longer crystal along the propagation direction, whereas in the longitudinal configuration Φ_L does not depend on l. Therefore the transverse configuration is a more desirable mode of operation. Besides, in the transverse case the field electrodes do not interfere with the incident light beam.

The final optical field passed by the output polarizer [along the $(\mathbf{a}_z - \mathbf{a}'_x)/\sqrt{2}$ direction] is

$$\frac{1}{\sqrt{2}}[E_z - E'_x] = \text{Re}\left\{\frac{E_0}{2}\exp\left(j\left[\omega_0 t - \frac{\omega_0}{c}n_e l\right]\right)\right.$$

$$\left. -\exp\left(j\left[\omega_0 t - \frac{\omega_0}{c}n'_x l\right]\right)\right\}. \quad (6.6\text{-}16)$$

Similar to the longitudinal case, the ratio of the output to the input intensities can be expressed in terms of the retardation as

$$\frac{I_0}{I_i} = \sin^2\left(\frac{\Phi_T}{2}\right). \quad (6.6\text{-}17)$$

However, note that in the transverse case, the output field is, in general, elliptically polarized in the absence of the applied field because, from Eq. (6.6-15), $\Phi_T = (\omega_0/c)l[n_o - n_e]$ at $V = 0$. This provides a means for changing linearly polarized light to elliptically polarized light electrooptically.

6.6.2 Phase Modulation

Figure 6.25 shows a system capable of phase-modulating the input light. Note that this system is similar to that of Figure 6.22. In amplitude modulation the input light is polarized along the bisector of the x' and y' axes, whereas in this *phase-modulation* scheme, the light is polarized parallel to one of the induced birefringence axes (x' in Figure 6.25). The applied electric field along z direction does not change the state of polarization, but only the phase. The phase shift at the end of the crystal is

$$\phi'_x = \frac{\omega_0}{c}n'_x l,$$

where n'_x is given by Eq. (6.6-5).

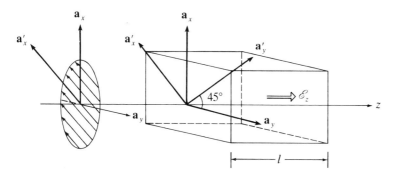

Figure 6.25 An electrooptic phase-modulating scheme.

If the applied field $\mathcal{E}_z$ is modulated according to

$$\mathcal{E}_z = \mathcal{E}_m \sin \omega_m t, \qquad (6.6\text{-}18)$$

the field at the exit of the crystal is

$$E_x' = \text{Re}\{E_0 \exp(j[\omega_0 t - \phi_x'])\}$$

$$= \text{Re}\left\{E_0 \exp(j\omega_0 t)\exp\left(\frac{-j\omega_0 n_o}{c}\right)\exp(-j\delta \sin \omega_m t)\right\}, \quad (6.6\text{-}19)$$

where

$$\delta = \frac{\omega_0 n_o^3 r_{63}\mathcal{E}_m l}{2c} = \frac{\pi n_o^3 r_{63}\mathcal{E}_m l}{\lambda_v} \qquad (6.6\text{-}20)$$

is called the *phase-modulation index*. The output light is phase-modulated. It is interesting to point out that the phase-modulation index is one-half the retardation Φ_L given by Eq. (6.6-9).

Problems

6.1 Light of wavelength λ_0 is incident on a grating of period d at an angle ϕ_{inc}, as shown in Figure P6.1. Show that the condition for maximum diffraction is

$$(\sin \phi_{\text{inc}} + \sin \theta)d = m\lambda_0, \qquad m = 0, 1, 2, \ldots.$$

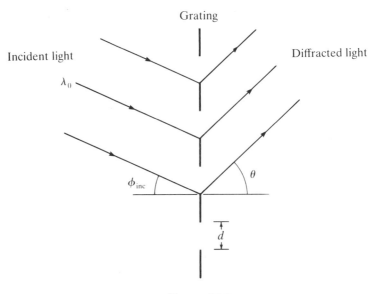

Figure P6.1

6.2 For oblique plane-wave incidence on the sound column of width L, that is, $\phi_{\text{inc}} \neq 0$, find the amplitude of the mth-order diffracted light. [*Hint*: Define a function

$$I_n = e^{jnbz} J_n[(a \sin bz)/b]$$

and show that $dI_n/dz = \frac{1}{2}ae^{2jbz}I_{n-1} - \frac{1}{2}ae^{-2jbz}I_{n+1}.]$

6.3 Starting from Eqs. (6.2-34) and (6.2-35) directly, show that

$$|\psi_0|^2 + |\psi_1|^2 = \text{constant}.$$

6.4 For *near phase-synchronous* downshifted acoustooptic Bragg diffraction, that is, $\phi_{\text{inc}} = \phi_B + \Delta\phi$, the zeroth and the negative first-order amplitudes evolve according to the following set of coupled equations:

$$\frac{d\psi_0}{dz} = -ja\psi_{-1}e^{-jKz\,\Delta\phi}$$

$$\frac{d\psi_{-1}}{dz} = -ja\psi_0 e^{+jKz\,\Delta\phi},$$

where a is an interaction constant and K denotes the sound wave number.

(a) Show that $|\psi_0|^2 + |\psi_{-1}|^2 = $ constant.

(b) Solve for ψ_0 and ψ_{-1} with the boundary conditions $\psi_0(z = 0) = \psi_{inc}$ and $\psi_{-1}(z = 0) = 0$.

6.5   As shown in Figure 6.11, the acoustooptic modulator is operated in the Bragg regime. The speed of sound in the acoustic medium is 4000 m/s, $\lambda_v = 0.6328$ μm, and the refraction index of the medium is 1.6.

     Two test signals from radio stations WTCP and WPPB at frequencies 40 MHz and 45 MHz are received by an antenna, amplified and fed to the transducer of the acoustooptic modulator. Assuming the focal length of the lens is 1 m, find the following:

(a) the frequencies of the light beams on the back focal plane of the lens;

(b) the spatial separation between the two beams in the back focal plane.

6.6   Given a beam of light that might be either unpolarized or circularly polarized, how might you determine its actual state of polarization?

6.7   A clockwise circularly polarized light ($\lambda_v = 0.688$ μm) described by

$$\mathbf{E}_i = E_i \cos(\omega_0 t - k_0 x)\mathbf{a}_y - E_i \sin(\omega_0 t - k_0 x)\mathbf{a}_z$$

is incident on an x-cut calcite crystal ($n_o = 1.658$, $n_e = 1.486$) of thickness $d = 0.006$ mm. Find the expression of the emergent $\mathbf{E}_{out}$ field and its state of polarization.

6.8   A clockwise circularly polarized light ($\lambda_v = 0.688$ μm) is incident on an x-cut quartz crystal ($n_o = 1.544$, $n_e = 1.553$) of thickness $d = 0.025$ mm. Determine the state of polarization of the emerging beam.

6.9   In the absence of an applied electric field, the ellipsoid for the uniaxial crystal $LiNbO_3$ is given by Eq. (6.5-1). An external electric field $\mathscr{E}_{z0}$ is now applied along the optic axis (z) of the crystal.

(a) Find the equation of the index ellipsoid and estimate the corresponding refractive indices along the x, y, and z directions.

(b) Assume that linearly polarized light is incident along the y axis, and the incident wave is decomposed into E_x and E_z components; find the phase retardation between the two components after the light traverses the length of the crystal, l.

(c) Calculate the required magnitude of $\mathscr{E}_{z0}$ if the emerging wave is circularly polarized. (Take $l = 1$ cm.)

References

6.1   Banerjee, P. P. and T.-C. Poon (1987). *Proc. Midwest Symp. Circuits and Systems* (G. Glasford and K. Jabbour, eds.) 820–823 North Holland, Amsterdam.

6.2   Fowles, G. R. (1975). *Introduction to Modern Optics*. Holt, Rinehart and Winston, New York.

6.3   Ghatak, A. (1977). *Optics*. Tata McGraw-Hill, New Delhi.

6.4   Haus, H. A. (1984). *Waves and Fields in Optoelectronics*. Prentice-Hall, Englewood Cliffs, New Jersey.

6.5   Hecht, E. and A. Zajac (1975). *Optics*. Addison-Wesley, Reading, Massachusetts.

6.6   Iizuka, K. (1985). *Engineering Optics*. Springer-Verlag, Berlin.

6.7   Johnson, C. C. (1965). *Field and Wave Electrodynamics*. McGraw-Hill, New York.

6.8   Korpel, A. (1972). Acousto-optics. In *Applied Solid State Science*, Vol. 3 (R. Wolfe, ed.) Academic, New York.

6.9   Korpel, A. (1988). *Acousto-Optics*. Marcel Dekker, Inc., New York and Basel.

6.10 Korpel, A., R. Adler, and B. Alpiner (1964). *Appl. Phys. Lett.* **5** 86.

6.11 Nussbaum, A. and R. A. Phillips (1976). *Contemporary Optics for Scientists and Engineers*. Prentice-Hill, Englewood Cliffs, New Jersey.

6.12 Pieper, R. J. and T.-C. Poon (1985). *IEEE Trans. Education* **E-28** 11.

6.13 Poon, T.-C. and S. K. Cheung (1989). *Appl. Optics* **28** 4787.

6.14 Poon, T.-C. and A. Korpel (1981). *J. Opt. Soc. Amer.* **71** 1202.

6.15 Whitman, R. L. and A. Korpel (1969). *Appl. Optics* **8** 1567.

6.16 Yariv, A. (1976). *Introduction to Optical Electronics*. Holt, Rinehart and Winston, New York.

Chapter 7   Lasers and Photodetectors

In all previous chapters, we discussed issues concerning optical propagation, information processing, and optical modulation, implicitly assuming that there exist sources of coherent light and measurement techniques to monitor optical intensity distributions whenever required. In this chapter, we will first discuss the basics of lasers, which are sources of coherent light. Specifically, we will present the concept of stimulated emission of coherent energy arising from the population inversion in the levels of an atomic system. We will then examine how the energy emitted builds up and is stored within an optical cavity or resonator. As an example of a practical system, we will present a simple description of the Helium–Neon (He–Ne) laser. In the second part of the chapter, we will introduce different types of devices capable of measuring optical radiation and will evaluate their performances in the presence of noise, in terms of the signal-to-noise ratio. We will also provide explicit examples of detection schemes that can be used in optical signal processing.

7.1 Spontaneous and Stimulated Emission

It is well known that atomic systems can exist in certain stationary states, each of which corresponds to a discrete value of the energy. These states are characterized by *quantum numbers* and the corresponding energy values $\tilde{E}_i$ are called the *levels* of the system [see, for instance, Messiah (1961)]. The lowest energy level

$(\tilde{E}_0)$ is termed the *ground level*. We point out that, sometimes, different states could have the same energy. This level is then called *degenerate*. For simplicity, we will begin by assuming a two-level system characterized by energy values $\tilde{E}_1$ and $\tilde{E}_2$, with $\tilde{E}_2 > \tilde{E}_1$, and *population densities* (i.e., the number of atoms per unit volume) N_1 and N_2, respectively. The system can make *transitions* between these two levels either by the *emission* of a photon of energy $\tilde{E} = \tilde{E}_2 - \tilde{E}_1 = \hbar\omega_0$, where ω_0 is the angular frequency of the radiation and $\hbar = h/2\pi$ (h is Planck's constant), or by *absorption* of a photon of the same energy. The emission can be *spontaneous* (i.e., may occur without any external excitation) or it can be *stimulated* by the presence of an external electromagnetic field.

(a) Spontaneous Emission

In this process, atoms in the level $\tilde{E}_2$ decay spontaneously to $\tilde{E}_1$ by losing energy in the form of radiation of a photon ($\hbar\omega$), which in general can travel in a random direction [see Figure 7.1(a)]. The rate of spontaneous emission R_{sp} from level $\tilde{E}_2$ can be written as

$$R_{sp} = A_{21}N_2 = -\frac{dN_2}{dt} \qquad (7.1\text{-}1)$$

if no other process takes place, where A_{21} is the *Einstein coefficient for spontaneous emission* and is indicative of the probability of emission that is independent of any external radiation density. Note that Eq. (7.1-1) reflects the fact that if N_2 decreases, N_1 must increase.

(b) Stimulated Emission

In this process, the atom in level $\tilde{E}_2$ gives up its energy $\hbar\omega$ at the same frequency, phase, polarization, and direction of the externally applied radiation [see Figure 7.1(b)] and goes to level $\tilde{E}_1$. The rate of stimulated emission R_{st} from level $\tilde{E}_2$ is given by

$$R_{st} = B_{21}N_2\rho(\omega), \qquad (7.1\text{-}2) \cdot$$

where B_{21} is the *Einstein coefficient for stimulated emission* and

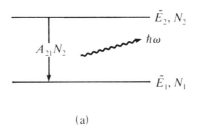

(a)

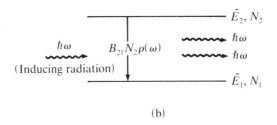

(b)

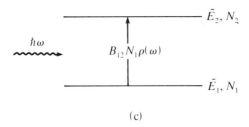

(c)

Figure 7.1 (a) Spontaneous emission. (b) Stimulated emission. (c) Absorption.

$\rho(\omega)\,d\omega$ is the electromagnetic radiation density (J/m^3) within the frequency interval ω to $\omega + d\omega$. Quantum mechanically, we can visualize the radiation field (inside a cavity) as a collection of an infinite number of simple harmonic oscillators, with each oscillator corresponding to a particular mode of the cavity. Using this analogy, the radiation density is given by *Planck's blackbody radiation law*,

$$\rho(\omega)\,d\omega = \frac{\hbar\omega^3}{\pi^2 v^3}\frac{1}{[\exp(\hbar\omega/k_B T)] - 1}\,d\omega, \qquad (7.1\text{-}3)$$

Lasers and Photodetectors

where k_B denotes the *Boltzmann constant* ($= 1.38 \times 10^{-23}$ J/°K), T is the temperature in kelvins, and v is the wave velocity in the medium.

(c) Absorption

Here, an atom in level $\tilde{E}_1$ absorbs a photon from the external field and goes to level $\tilde{E}_2$ [see Figure 7.1(c)]. The rate of absorption R_{abs} in level $\tilde{E}_1$ is

$$R_{abs} = B_{12}N_1\rho(\omega), \tag{7.1-4}$$

where B_{12} is the *Einstein coefficient for absorption* and $\rho(\omega)$ is as defined in Eq. (7.1-3). The reason we use the same $\rho(\omega)$ in both Eqs. (7.1-2) and (7.1-4) is that the same external radiation causes stimulated emission from $\tilde{E}_2$ and absorption in $\tilde{E}_1$.

At thermal equilibrium, the number of upward transitions ($\tilde{E}_1 \to \tilde{E}_2$) is equal to the number of downward transitions ($\tilde{E}_2 \to \tilde{E}_1$). Thus, combining Eqs. (7.1-1), (7.1-2), and (7.1-4), we have

$$(B_{12}N_1 - B_{21}N_2)\rho(\omega) - A_{21}N_2 = 0. \tag{7.1-5a}$$

or

$$\rho(\omega) = \frac{A_{21}N_2}{B_{12}N_1 - B_{21}N_2}. \tag{7.1-5b}$$

Now, employing *Boltzmann statistics* [see, for instance, Kittel (1976)], the number density of states N_1 and N_2 having degeneracies $\tilde{g}_1$ and $\tilde{g}_2$, respectively, are related by

$$\frac{N_2}{\tilde{g}_2} = \frac{N_1}{\tilde{g}_1}\exp\left(\frac{-\hbar\omega}{k_BT}\right) = \frac{N_1}{\tilde{g}_1}\exp\left(\frac{-(\tilde{E}_2 - \tilde{E}_1)}{k_BT}\right), \tag{7.1-6}$$

where $\tilde{g}_i$ denotes the number of states in the degenerate level $\tilde{E}_i$. Incorporating Eq. (7.1-6) into Eq. (7.1-5b), we get

$$\rho(\omega) = \frac{A_{21}}{(\tilde{g}_1/\tilde{g}_2)\exp(\hbar\omega/k_BT)B_{12} - B_{21}}, \tag{7.1-7}$$

which has the same functional form as Eq. (7.1-3)! Indeed, a comparison of Eq. (7.1-3) with Eq. (7.1-7) suggests that

$$\frac{A_{21}}{B_{21}} = \frac{\hbar \omega^3}{\pi^2 v^3} \qquad (7.1\text{-}8)$$

and

$$\tilde{g}_1 B_{12} = \tilde{g}_2 B_{21}. \qquad (7.1\text{-}9)$$

Equations (7.1-8) and (7.1-9) relate the Einstein coefficients A_{21}, B_{21}, and B_{12} in thermodynamic equilibrium. Note that if the system is nondegenerate, $\tilde{g}_1 = \tilde{g}_2 = 1$ and hence $B_{12} = B_{21}$. We mention at this point that A_{21} does not contribute to photon multiplication resulting in a coherent laser output. B_{12}, however, contributes to stimulated coherent emission.

Based on the preceding analysis, we can advance the following remarks:

Remark 1: Observe, from Eq. (7.1-8), that $A_{21}/B_{21} \propto \omega^3$. Hence, as the frequency increases, so does the amount of spontaneous emission, resulting in proportionally less coherent output. This suggests that lasers of higher frequency, such as in the ultraviolet or X-ray, are more difficult to build and operate.

Remark 2: The ratio of the rates of depletion of level $\tilde{E}_2$ due to spontaneous and stimulated emission [see Eqs. (7.1-1)–(7.1-3)] is

$$\frac{R_{sp}}{R_{st}} = \left(\frac{A_{21}'}{B_{21}}\right)\frac{\pi^2 v^3}{\hbar \omega^3}\left(\exp\left[\frac{\hbar \omega}{k_B T}\right] - 1\right) \propto \exp\left(\frac{\hbar \omega}{k_B T}\right) - 1. \quad (7.1\text{-}10)$$

For an optical source at frequency $\omega \sim 2\pi \times 5 \times 10^{14}$ rad/s operating at $T \sim 500°K$, it follows that $R_{sp} \gg R_{st}$. Thus, under normal circumstances, the output from light sources is incoherent.

Remark 3: Amplification of coherent light may occur even under the preceding circumstances if the rate of stimulated emission exceeds the absorption rate. The difference between these rates, say, for a nondegenerate atom, is [using Eqs. (7.1-2) and (7.1-4)]

$$R_{st} - R_{abs} = B_{12}\rho(\omega)(N_2 - N_1). \qquad (7.1\text{-}11)$$

Normally, assuming Boltzmann statistics, $N_2 < N_1$ and there is no amplification. However, if $N_2 > N_1$, (which is achievable through techniques that we shall mention later), we have *population inversion*, and hence, amplification of the stimulated emission, resulting in coherent light. This will be discussed in more detail in the following section.

7.2 Amplification through Population Inversion in a Gain Medium

In the previous section, we examined stimulated emission and absorption in a two-level system under external radiation. Consider a slab of a material that is a two-level system, as shown in Figure 7.2. For a given input intensity, we will find the evolution of the intensity along the propagation direction (z) of the radiation. Physically, two competing processes (i.e., stimulated emission and absorption) govern this intensity evolution. To achieve a more physical model, we will no longer assume discreteness in levels $\tilde{E}_1$ and $\tilde{E}_2$ but will allow for number densities of atoms in these levels to be a function of frequency. Therefore, let $\tilde{n}_1(\omega)\,d\omega$ and $\tilde{n}_2(\omega)\,d\omega$ represent the number of atoms per unit volume in levels $\tilde{E}_1$ and $\tilde{E}_2$ that are capable of absorbing radiation in a bandwidth $d\omega$ around ω and of stimulated emission in the same bandwidth, respectively. Referring to the volume element in Figure 7.2, the requirement of continuity demands that the net increase in the intensity of radiation must be proportional to the net rate at which photons are generated within the volume. Let $I(z;\omega)\,d\omega$ represent the energy through a unit area per unit time whose frequencies lie between ω and $\omega + d\omega$. Then the preceding requirement translates to

$$[I(z + dz;\omega)\,d\omega - I(z;\omega)\,d\omega]S$$

$$= S\hbar\omega B[\tilde{n}_2(\omega)\,d\omega - \tilde{n}_1(\omega)\,d\omega]\rho(z;\omega)\,dz,$$

Amplification

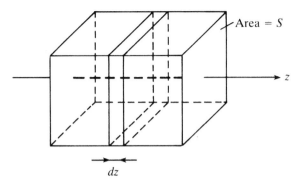

Figure 7.2 Light propagating along the z-direction through the laser medium.

implying

$$\frac{\partial I(z;\omega)}{\partial z} = \hbar\omega B\rho(z;\omega)[\tilde{n}_2(\omega) - \tilde{n}_1(\omega)], \qquad (7.2\text{-}1)$$

where $\rho(z;\omega)\,d\omega$ denotes the *radiation density* at z. In obtaining Eq. (7.2-1), we have assumed a nondegenerate two-level system so that

$$B_{12} = B_{21} = B. \qquad (7.2\text{-}2)$$

Now, $I\,d\omega$ is related to $\rho\,d\omega$ through

$$I(z;\omega)\,d\omega = \upsilon\rho(z;\omega). \qquad (7.2\text{-}3)$$

Combining Eqs. (7.2-1) and (7.2-3), we obtain

$$\frac{\partial I(z;\omega)}{\partial z} = -\alpha(\omega)I(z;\omega), \qquad (7.2\text{-}4)$$

where $\alpha(\omega)$ is an *absorption coefficient* defined as

$$\alpha(\omega) = \frac{\hbar\omega}{\upsilon}[\tilde{n}_2(\omega) - \tilde{n}_1(\omega)]B. \qquad (7.2\text{-}5)$$

Equation (7.2-4) with initial conditions $I(0; \omega)$ has the solution

$$I(z; \omega) = I(0; \omega)e^{-\alpha(\omega)z}. \tag{7.2-6}$$

In principle, we can find the total intensity at a certain z due to all possible frequencies by directly integrating Eq. (7.2-6) with respect to ω. However, we will confine ourselves to exploring the nature of $\alpha(\omega)$. Thus, we integrate Eq. (7.2-4) to get

$$\int \frac{1}{I} \frac{\partial I}{\partial z} d\omega = -\int \alpha(\omega) d\omega$$

$$= \frac{\hbar}{v} \int \omega B [\tilde{n}_2(\omega) - \tilde{n}_1(\omega)] d\omega. \tag{7.2-7}$$

Because $\tilde{n}_1(\omega)$ and $\tilde{n}_2(\omega)$, and hence $\alpha(\omega)$, are sharply peaked functions around $\omega = \omega_0$ (their nature will be discussed later), we write, for simplicity,

$$\tilde{n}_1(\omega) = N_1 \delta(\omega - \omega_0), \qquad \tilde{n}_2(\omega) = N_2 \delta(\omega - \omega_0). \tag{7.2-8}$$

Then, from Eq. (7.2-7),

$$\int \frac{1}{I} \frac{\partial I}{\partial z} d\omega = \kappa(N_2 - N_1), \tag{7.2-9}$$

where, using Eq. (7.1-8),

$$\kappa = \frac{\hbar \omega_0 B}{v} = \frac{A_{21} \pi^2 v^2}{\omega_0^2}. \tag{7.2-10}$$

In general, we can write

$$\alpha(\omega) = \kappa(N_1 - N_2)g(\omega) \tag{7.2-11}$$

with the function $g(\omega)$ properly normalized so that

$$\int g(\omega) d\omega = 1. \tag{7.2-12}$$

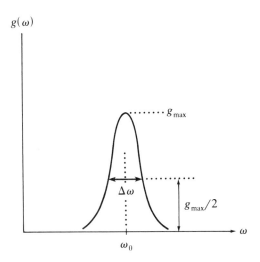

Figure 7.3 Typical line-shape function $g(\omega)$ vs. ω.

The function $g(\omega)$ is called the *line-shape function*. A typical plot of $g(\omega)$ is shown in Figure 7.3. In general, there are two mechanisms that contribute to the width $\Delta\omega$ of the line. They are called homogeneous and inhomogeneous broadening and will be discussed in more detail later.

We return, for a short while, to a discussion of Eq. (7.2-6) with Eq. (7.2-11). At thermal equilibrium, the population N_1 of the ground state is greater than that (N_2) of the excited state; hence, the intensity I is attenuated as light propagates through the medium. However, if there is population inversion (i.e., $N_2 > N_1$), the light is amplified. Population inversion in a laser medium may be produced optically, electrically, chemically, or thermally by an external energy source, or *pump*. For gas lasers, such as the He–Ne laser to be discussed later, the most commonly used pump is an electrical discharge.

From Eq. (7.1-1), note that the rate of change of population at level $\tilde{E}_2$ due to spontaneous emission is related to the population density of $\tilde{E}_2$ by the proportionality constant $-A_{21}$. This implies that the population density exponentially decreases with a time constant $\tau = 1/A_{21}$. For efficient laser operation, we must ensure that the population inversion, once created through pumping, persists for a long time and does not decay due to spontaneous

Lasers and Photodetectors

emission. This implies that τ must be made large, or A_{21} must be small. This means that the excited level should be in a long-lifetime, or *metastable*, state. At the same time, note from Eqs. (7.2-6) and (7.2-11), that a smaller A_{21} means that the stimulated emission takes a longer distance of propagation to grow considerably.

We now return to a discussion of the mechanisms that cause broadening of the line-shape functions. The idea of a line-shape function is most important because it applies to all three processes (i.e., spontaneous emission, stimulated emission, and absorption) that we defined in connection with the two-level system. One of the contributions to *homogeneous broadening* is the finite lifetime $\tau = 1/A_{21}$ of the emitting state during spontaneous emission. We can model this emission using a damped oscillator model. Then the emitted optical field as a function of time is given by

$$E(z,t) \propto e^{-t/\tau} \cos \omega_0 t \quad \text{for } t \geq 0. \qquad (7.2\text{-}13a)$$

The power spectrum corresponding to the electric field [indicative of $I(z;\omega)$] is

$$|F_t\{E(z,t)\}|^2 \propto \frac{1}{(1/\tau)^2 + (\omega - \omega_0)^2} \propto g(\omega), \qquad (7.2\text{-}13b)$$

with $1/\tau \ll \omega_0$ and where we assume $\omega \sim \omega_0$. In deriving (7.2-13b), the definition

$$F_t\{f(t)\} \triangleq \int_{-\infty}^{\infty} f(t)e^{-j\omega t} \, dt \qquad (7.2\text{-}14)$$

(see Chapter 1) has been employed. Defining $\Delta\omega_h = 2/\tau$ and normalizing Eq. (7.2-13b) according to $\int g(\omega) \, d\omega = 1$, we obtain the normalized line-shape function,

$$g(\omega) = \frac{\Delta\omega_h}{(\omega - \omega_0)^2 + (\Delta\omega_h/2)^2}, \qquad (7.2\text{-}15)$$

which has the *Lorentzian* form, the name originating from the Lorentz transformation of electromagnetic fields. Another contribution to homogeneous broadening is from *collisions*. In a gas, it is due to the collision of an atom with other atoms, ions and so forth; in a solid it is due to the interaction of the atom with the lattice. This mechanism also gives a Lorentzian line-shape function whose width depends on the average time between collisions. The typical range of $\Delta \omega_h / 2\pi$ is from 0.5–20 MHz in lasers.

Note that, in the preceding discussion, there is no distinction made between one type of atom and another, that is, there is no assumed distinguishing feature about any group of atoms. This is characteristic of homogeneous broadening. On the other hand, *inhomogeneous broadening* occurs due to the presence of a distinguishing feature of a certain group of atoms. For instance, in naturally occurring neon, 80% has an atomic mass of 20, whereas the rest has the atomic mass of an isotope, namely, 22. This causes an asymmetric line shape referred to as the *isotope shift*.

Another contribution to inhomogeneous broadening is the thermal motion of some gaseous atoms and is called *Doppler broadening*. When a moving atom interacts with radiation, the apparent frequency of the incident wave is different from the frequency observed from a stationary atom. Therefore, the resonant frequency of the atom is Doppler-shifted and is given by

$$\omega = \omega_0 - \frac{V_z}{v}\omega_0, \qquad v \gg V_z, \qquad (7.2\text{-}16)$$

where ω is the apparent resonant frequency, ω_0 is the transition frequency of the atom, and V_z is the relative component velocity in the direction of the axis of the laser. In order to find the line-shape function due to this phenomenon, we note that the Maxwell velocity distribution function of an atom of mass M with velocity components V_x, V_y, V_z in a gas at temperature T is

$$p(V_x, V_y, V_z) = p(V_x)p(V_y)p(V_z)$$

$$= \left(\frac{M}{2\pi kT}\right)^{3/2} \exp\left[-\frac{M(V_x^2 + V_y^2 + V_z^2)}{2k_B T}\right]. \quad (7.2\text{-}17)$$

Because z is the axis of the laser, integration over the x and y velocity components yields

$$p(V_z)\,dV_z = \left(\frac{M}{2\pi k_B T}\right)^{1/2} \exp\left(-\frac{MV_z^2}{2k_B T}\right)\,dV_z, \quad (7.2\text{-}18)$$

which represents the probability that an atom has a z-component of velocity lying between V_z and $V_z + dV_z$. Therefore, the desired probability $g(\omega)\, d\omega$ that the transition frequency is between ω and $\omega + d\omega$ can be found by substituting Eq. (7.2-16) into Eq. (7.2-18),

$$g(\omega)\, d\omega = \left(\frac{M}{2\pi k_B T}\right)^{1/2} \exp\left[-\frac{M}{2k_B T}\frac{v^2}{\omega_0^2}(\omega - \omega_0)^2\right]\frac{v}{\omega_0}\, d\omega.$$

$$(7.2\text{-}19)$$

$g(\omega)$ is a Gaussian distribution and its full width at half-maximum (FWHM) is

$$\Delta\omega_D = 2\omega_0\sqrt{\frac{2k_B T \ln 2}{Mv^2}}.\qquad (7.2\text{-}20)$$

Combining Eq. (7.2-19) with Eq. (7.2-20), we have the normalized Gaussian distribution in terms of $\Delta\omega_D$,

$$g(\omega) = \frac{4\pi}{\Delta\omega_D}\sqrt{\frac{\ln 2}{\pi}}\exp\left(-\left[\frac{\omega - \omega_0}{\Delta\omega_D/2}\right]^2 \ln 2\right).\qquad (7.2\text{-}21)$$

In general, a Gaussian broadening will result from any mechanism producing a random distribution of the transition frequencies of the atoms. Other inhomogeneous broadening mechanisms, such as the imperfections of the atomic lattice that cause local variations of the energy levels of the atom, also give a *Gaussian line-shape function*. A typical inhomogeneous-broadening line-width is of the order of gigahertz.

7.3 Three- and Four-Level Systems

In Sections 7.1 and 7.2 we studied population inversion and gain mechanism in a simple two-level system. In actuality, lasers are better modelled as three- or four-level systems. The ruby lasers can be accurately represented as a three-level system, whereas He–Ne, ion and CO_2 lasers are four-level systems. In this section, we will briefly describe those systems where population inversion is achieved between two levels by making use of other energy levels.

Consider an idealized three-level laser system as shown in Figure 7.4. Energy from an appropriate pump is coupled into the laser medium. The pump lifts atoms from level 1 ($\tilde{E}_1$) into several excited states, collectively labeled $\tilde{E}_3$. It is important that level 3 is a broad level because a broadband optical source then can be used efficiently as a pumping source. From these states, the atoms decay spontaneously to level 2 ($\tilde{E}_2$). This decay is usually radiationless, and the energy is given to the surrounding molecules in gas lasers and to the lattice of atoms in solids. Because level 2 is metastable, atoms funnel rapidly from pump level $\tilde{E}_3$ to level $\tilde{E}_2$ and begin to pile up at the metastable level. The consequence is population inversion in level 2, which is required for light amplification by stimulated emission. Initiated by spontaneous emission, lasing can then occur between levels 2 and 1.

In idealized four-level lasers, as shown in Figure 7.5, level 0 is the ground level and levels 1, 2, and 3 are excited levels of the system. Atoms from level 0 are pumped to level 3, from which the atoms decay nonradiatively to the metastable level 2, with lasing occurring between that level (usually called the *upper laser level*) and another intermediate level (the *lower laser level*) $\tilde{E}_1$. The lower laser level is an ordinary level that decays rapidly to ground level $\tilde{E}_0$ by nonradiative transition, so that the population in level 1 cannot build up to a large value. The overall effect is then population inversion between levels 2 and 1.

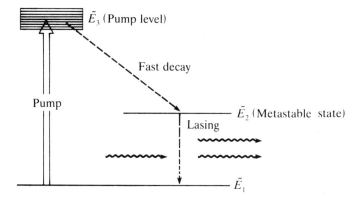

Figure 7.4 Three-level laser scheme.

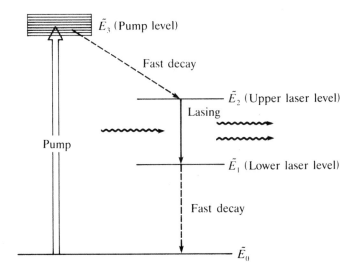

Figure 7.5 Four-level laser scheme.

7.4 Optical Resonators and Hermite – Gaussian Beams

We have studied how a light beam passing through an active (laser) medium with population inversion can be amplified. In order to construct an oscillator from an amplifier, it is necessary to introduce a suitable positive feedback. The feedback is often provided by placing the amplifying medium between two mirrors, that is, by placing the medium inside an optical resonator. In fact, optical resonators form an integral part of a laser system. An optical resonator can confine and direct a growing number of photons, created through stimulated emission, back and forth through the laser medium, continuously exploiting the population inversion to create more and more stimulated emissions. In order to couple a part of the laser light out of the resonator, one of its mirrors is made partially reflecting.

In Chapter 2, we analyzed an optical resonator bounded by two spherical mirrors using geometrical optics. In this section, we will perform a wave optics analysis to find explicitly the beam profiles of different *modes* in the resonator, and hence the field

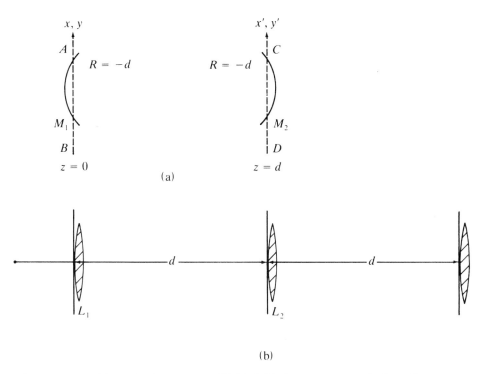

Figure 7.6 (a) Optical resonator. (b) Unfolded resonator equivalent diagram using lenses.

distributions of the output light beam. For simplicity, we consider a *confocal resonator system* (see Section 2.4.4) that consists of a pair of concave mirrors of equal radii of curvature $R \, (= -d)$ separated by a distance d. The resonator system is shown in Figure 7.6(a). In order to calculate the transverse modes inside the resonator, we let $\psi_p(x, y)$ represent the optical field along the plane AB as shown in Figure 7.6(a). This field $\psi_p(x, y)$ undergoes Fresnel diffraction over a distance d, is then reflected from mirror M_2, travels back to mirror M_1, and undergoes a second reflection in one round trip; $\psi_p(x, y)$ is called a mode of the resonator if it reproduces itself after one round trip (apart from some constant). This, in turn, implies that, besides other requirements, the phase fronts of $\psi_p(x, y)$ coincide with the phase fronts of the field after one round

Lasers and Photodetectors

trip. From geometry, it is easy to see that the phase fronts along planes AB and CD (which are separated by d) are symmetric about $z = d/2$.

The resonator in Figure 7.6(a) may be "unfolded" as shown in Figure 7.6(b). In this equivalent model, the "unfolding" implies that the phase fronts immediately behind the lens L_2 are identical to the phase fronts immediately behind lens L_1. The focal length f of each lens must be equal to $-R/2 = d/2$. Now, if $\psi_p(x, y)$ and $\tilde{\psi}_p(x, y)$ represent the optical fields immediately behind lenses L_1 and L_2, respectively, we must impose

$$\tilde{\psi}_p(x, y) = \gamma \psi_p(x, y), \qquad (7.4\text{-}1)$$

where γ is some complex constant. Explicitly, Eq. (7.4-1) reads [see Eq. (3.3-19)]:

$$\gamma \psi_p(x, y) = \left[\frac{jk_0}{2\pi d} \exp(-jk_0 d) \iint_S \psi_p(x', y') \right.$$

$$\times \exp\left(-j\frac{k_0}{2d}\left[(x - x')^2 + (y - y')^2 \right] \right) dx'\, dy' \Bigg]$$

$$\times \exp\left[j\frac{k_0}{d}(x^2 + y^2) \right]$$

$$= j\frac{k_0 \exp(-jk_0 d)}{2\pi d} \exp\left[j\frac{k_0}{2d}(x^2 + y^2) \right] \iint_S \psi_p(x', y')$$

$$\times \exp\left[-j\frac{k_0}{2d}(x'^2 + y'^2 - 2xx' - 2yy') \right] dx'\, dy', \quad (7.4\text{-}2)$$

where S is the area of mirrors M_1 and M_2 in Figure 7.6(a). Equation (7.4-2) is an integral equation with $\psi_p(x, y)$ to be solved.

As we will see shortly, there are an infinite number of solutions ψ_{mn} called *eigenfunctions*, or *eigenmodes*, each with an associated *eigenvalue* γ_{mn}, where m and n denote the *transverse mode numbers*. The mode numbers determine the transverse field distribution of the mode. The eigenvalues γ_{mn} have physical meanings. If we put

$$\gamma_{mn} = |\gamma_{mn}| e^{j\phi_{mn}}, \qquad (7.4\text{-}3)$$

we see that the quantity $1 - |\gamma_{mn}|^2$ gives the energy loss per half-cycle transit. The loss is called the diffraction loss of the resonator and is due to the energy "spills" around the reflecting mirrors [because we are considering the so-called open-sided resonators, see Figure 7.6(a), in which we assume that S is finite]. The phase of γ_{mn} (i.e., ϕ_{mn}) is the phase shift per half-cycle transit, which determines the oscillation frequencies of the resonator.

Returning to Eq. (7.4-2) and defining a function $f(x, y)$ through the equation

$$f(x, y) = \psi_p(x, y) \exp\left[-j\frac{k_0}{2d}(x^2 + y^2)\right], \qquad (7.4\text{-}4)$$

Eq. (7.4-2) becomes

$$\gamma f(x, y) = \frac{jk_0 \exp(-jk_0 d)}{2\pi d} \int_{-a}^{a} \int_{-a}^{a} f(x', y')$$

$$\times \exp\left[j\frac{k_0}{d}(xx' + yy')\right] dx' dy', \qquad (7.4\text{-}5)$$

where we have assumed that the two mirrors have square cross sections of linear dimension 2a.

Introducing the dimensionless variables ξ and η through the relations

$$\xi = \sqrt{2\pi N}\,\frac{x}{a} \quad \text{and} \quad \eta = \sqrt{2\pi N}\,\frac{y}{a}, \qquad (7.4\text{-}6)$$

where $N \triangleq a^2 k_0/2\pi d$ is called the Fresnel number of the square aperture, Eq. (7.4-5) reduces to

$$\gamma f(\xi, \eta) = \frac{je^{-jk_0 d}}{2\pi} \int_{-\sqrt{2\pi N}}^{\sqrt{2\pi N}} \int_{-\sqrt{2\pi N}}^{\sqrt{2\pi N}} f(\xi', \eta') e^{j(\xi\xi' + \eta\eta')} d\xi' d\eta'.$$

$$(7.4\text{-}7)$$

To solve for $f(\xi, \eta)$, we use the separation of variables technique and write

$$\gamma = \gamma_1 \gamma_2,$$

$$f(\xi, \eta) = p(\xi) q(\eta).$$

(7.4-8)

On substitution, Eq. (7.4-7) can be written as two separate equations,

$$\gamma_1 p(\xi) = \exp\left(-\frac{jk_0 d}{2}\right) \sqrt{\frac{j}{2\pi}} \int_{-\sqrt{2\pi N}}^{\sqrt{2\pi N}} p(\xi') \exp(j\xi\xi') \, d\xi' \quad (7.4\text{-}9)$$

and

$$\gamma_2 q(\eta) = \exp\left(-\frac{jk_0 d}{2}\right) \sqrt{\frac{j}{2\pi}} \int_{-\sqrt{2\pi N}}^{\sqrt{2\pi N}} q(\eta') \exp(j\eta\eta') \, d\eta'.$$

(7.4-10)

The solutions of these integral equations are *prolate spheroidal functions* [Slepian and Pollack (1961)] and these functions are numerically tabulated. Here we will only consider the case when $N \gg 1$. For such a case, we can extend the limits of integration in Eqs. (7.4-9) and (7.4-10) from $-\infty$ to ∞. Equations (7.4-9) and (7.4-10) then tell us that the functions $p(\xi)$ and $q(\eta)$ are their own Fourier transforms apart from some constant. A complete set of functions that satisfy the condition is the Hermite–Gaussian functions.

In Chapter 3, we have already been exposed to the Hermite–Gaussian functions. Recall that they were introduced via the underlying ODE. This time, for variety, we present an integral form that defines the Hermite–Gaussian functions:

$$j^m g_m(\xi) = \frac{1}{\sqrt{2\pi}} \int_{-\infty}^{\infty} g_m(\xi') e^{j\xi\xi'} \, d\xi', \qquad (7.4\text{-}11)$$

where $g_m(\xi) \triangleq H_m(\xi) e^{-\xi^2/2}$ and $H_m(\xi)$ is the Hermite polynomial of order m, defined by

$$H_m(\xi) \triangleq (-1)^m e^{\xi^2} \frac{d^m}{d\xi^m} e^{-\xi^2}. \qquad (7.4\text{-}12)$$

Hence, in order that $p(\xi)$ and $q(\eta)$ be of the Hermite–Gaussian form, it is required that

$$\gamma_1 = j^{(m+1/2)} \exp\left(-\frac{jk_0 d}{2}\right) = \exp\left\{-j\left[\frac{1}{2}k_0 d - \frac{1}{2}\left(m + \frac{1}{2}\right)\pi\right]\right\}$$

$$(7.4\text{-}13a)$$

and

$$\gamma_2 = j^{(n+1/2)} \exp\left(-\frac{jk_0 d}{2}\right) = \exp\left\{-j\left[\frac{1}{2}k_0 d - \frac{1}{2}\left(n + \frac{1}{2}\right)\pi\right]\right\}.$$

$$(7.4\text{-}13b)$$

The solutions of Eqs. (7.4-9) and (7.4-10) are, therefore, the Hermite–Gaussian functions

$$p_m(\xi) = H_m(\xi) e^{-\xi^2/2} \qquad (7.4\text{-}14a)$$

and

$$q_n(\eta) = H_n(\eta) e^{-\eta^2/2}, \qquad (7.4\text{-}14b)$$

where m and n denote the transverse mode numbers and determine the field distribution of the mode. Thus the complete solution of Eq. (7.4-2) with $N \gg 1$ can be expressed as *Hermite–Gaussian beams*,

$$\psi_{p_{mn}}(x, y) = A \exp\left[j\frac{1}{2}(\xi^2 + \eta^2)\right] p_m(\xi) q_n(\eta) \Big|_{\substack{\xi = \sqrt{2\pi N} x/a \\ \eta = \sqrt{2\pi N} y/a}}$$

$$= A \exp\left[j\frac{k_0}{2d}(x^2 + y^2)\right] H_m\left(\sqrt{\frac{k_0}{d}}\, x\right) H_n\left(\sqrt{\frac{k_0}{d}}\, y\right)$$

$$\times \exp\left[-\frac{k_0}{2d}(x^2 + y^2)\right], \qquad (7.4\text{-}15)$$

where A is some constant and we have made use of Eqs. (7.4-4), (7.4-8), and (7.4-14). Each set (m, n) corresponds to a particular transverse electromagnetic (TEM) mode of the resonator as the electric (and magnetic) field of the electromagnetic wave is orthogonal to the resonator z axis. The lowest-order Hermite polynomial H_0 is equal to unity; hence, the mode corresponding to the set $(0, 0)$ is called the TEM_{00} mode and has a Gaussian radial profile. Figure 3.22 shows the three lowest-order Hermite–Gaussian functions. Note that, in general, the mth-order mode contains m nulls (or zeros) and $m - 1$ peaks. Figure 3.23 depicts the intensity patterns of some low-order modes. Note that higher-order modes have a more spread-out intensity distribution and therefore would have higher diffraction losses. Most practical lasers are made to oscillate in the TEM_{00} mode. Because higher-order modes have wide transverse dimensions, they can be suppressed by placing a circular aperture inside the resonator.

So far we have discussed only the eigenfunctions of Eqs. (7.4-9) and (7.4-10). In discussing their corresponding eigenvalues, we note that from Eq. (7.4-13),

$$\gamma = \gamma_1 \gamma_2 = \exp\left\{ -j\left[k_0 d - (m + n + 1)\frac{\pi}{2}\right] \right\}. \quad (7.4\text{-}16)$$

Observe that $|\gamma| = 1$, implying that the diffraction losses are zero. This is expected, as in our analysis we have essentially assumed the mirror cross-section is of extremely large dimensions, that is, $N \gg 1$.

We have mentioned previously that the phase of γ represents the additional phase shift per half-cycle transit. The condition for supporting a mode is that the field does not change in a round trip through the resonator. This means that the phase change of the field in a round trip should be an integral multiple of 2π, or for the half-cycle transit, the phase change must be an even multiple of π. Therefore, we must have

$$k_0 d - (m + n + 1)\frac{\pi}{2} = l\pi, \qquad l = 1, 2, 3, \ldots, \quad (7.4\text{-}17)$$

where l refers to the *longitudinal mode number*.

Hence, the *resonant frequencies* of the oscillation of the resonator are expressible as

$$\omega_{lmn} = \pi(2l + m + n + 1)\frac{v}{2d},\qquad(7.4\text{-}18)$$

and only those frequencies that satisfy this equation are allowed in the resonator. Note that modes having the same value of $2l + m + n + 1$ have the same resonance frequency, although they have different field distributions. These modes are therefore *degenerate*. The frequency spacing between two modes with the same value of m and n and with l differing by 1 is

$$\omega_{l+1} - \omega_l = \Delta\omega_l = \frac{\pi v}{d},\qquad(7.4\text{-}19)$$

and the frequency separation between two transverse modes having the same value of l is

$$\Delta\omega_m = \frac{\pi v}{2d} = \Delta\omega_n.\qquad(7.4\text{-}20)$$

The field distribution Eq. (7.4-15) represents the field along the plane AB as shown in Figure 7.6(a), and the field distribution in the plane midway between the two mirrors can be evaluated, using the Fresnel integral, as

$$\psi_{P_{mn}}\Big|_{z=d/2} = jA\exp\left(-\frac{jk_0 d}{2}\right)\exp\left[-\frac{(x^2 + y^2)}{w_0^2}\right]$$

$$\times H_m\left(\frac{\sqrt{2}\,x}{w_0}\right)H_n\left(\frac{\sqrt{2}\,y}{w_0}\right),\qquad(7.4\text{-}21)$$

where $w_0 = \sqrt{d/k_0}$ is called the *waist* of the beam. Note that the beam has no phase curvature. Therefore, the phase fronts are planar midway between the mirrors. Also, from Eq. (7.4-15) note that the phase fronts of the field distribution at the mirrors M_1 and M_2 have radii of curvature that are identical to the radii of the mirrors.

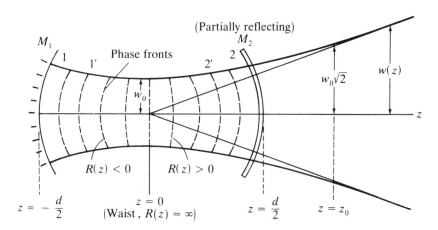

Figure 7.7 Confocal resonator system.

In practice, the laser output comprises a small fraction of the energy in the resonator that is coupled out through the mirror M_2, which is made partially reflective. We will be interested to know the behavior of the laser output with propagation. To find this, it is convenient to have the resonator center taken to the origin $z = 0$, so that the field at $z = 0$ (see Figure 7.7) can be expressed as

$$\psi_{p_{mn}}(x, y, z = 0) = E_0 \exp\left[-\frac{(x^2 + y^2)}{w_0^2}\right] H_m\left(\frac{\sqrt{2}\,x}{w_0}\right) H_n\left(\frac{\sqrt{2}\,y}{w_0}\right).$$

$$(7.4-22)$$

This is a more general form of the Gaussian initial profile discussed in Chapter 3. One can employ the Fresnel integral formula to find the field at any plane z inside as well as outside the resonator:

$$\psi_{p_{mn}}(x, y, z) = \frac{E_0 w_0}{w(z)} \exp(-jk_0 z)\exp\left[-\frac{jk_0(x^2 + y^2)}{2R(z)}\right]$$

$$\times \exp[-j(m + n + 1)\phi(z)]$$

$$\times H_m\left(\frac{\sqrt{2}\,x}{w(z)}\right) H_n\left(\frac{\sqrt{2}\,y}{w(z)}\right)\exp\left[-\frac{(x^2 + y^2)}{w^2(z)}\right], \quad (7.4-23)$$

where $w(z)$, $\phi(z)$, $R(z)$, and w_0 are as defined in Eqs. (3.3-34)–(3.3-36) with $z_R = d/2$ defined in connection with the diffraction of Gaussian beams. We remark that the resonant frequencies can also be calculated through the function $\phi(z)$. In fact, by evaluating the phase term, namely, $k_0 z + (m + n + 1)\phi(z)$, at $z = \pm d/2$, and setting the difference equal to $l\pi$, we have

$$k_{lmn}d + (m + n + 1)\left[\phi\left(\frac{d}{2}\right) - \phi\left(-\frac{d}{2}\right)\right] = l\pi, \qquad k_{lmn} = \frac{\omega_{lmn}}{v},$$

$$(7.4\text{-}24)$$

which is identical to Eq. (7.4-18).

Example 7.1 Stable Two-Mirror Resonators

We consider a resonator, as shown in Figure 7.8, which consists of the two mirrors of radii of curvature R_1 and R_2 separated by a distance d. Note that R_1 and R_2 are both negative according to our sign convention. We shall determine the parameters of the Gaussian beams that will just fit properly between these two mirrors. We assume the Gaussian beam has an initially unknown beam waist w_0 at $z = 0$. The essential conditions in order

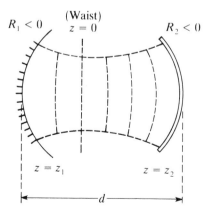

Figure 7.8 A resonator formed of two spherical mirrors of radii R_1 and R_2 and separated by a distance d. The waist is at $z = 0$.

that the Gaussian beam may keep bouncing back and forth between the mirrors are that the wave-front curvature $R(z)$ of the beam must match the mirror curvature at each mirror. This provides us with two equations [see Eq. (3.3-35)],

$$R(z_1) = z_1 \left[1 + \left(\frac{z_R}{z_1} \right)^2 \right] = +R_1,$$

$$\tag{7.4-25}$$

$$R(z_2) = z_2 \left[1 + \left(\frac{z_R}{z_2} \right)^2 \right] = -R_2,$$

where $z_R = k_0 w_0^2 / 2$ is the Rayleigh range, as defined in Example 3.6 (see Figure 7.7). Also,

$$z_2 - z_1 = d. \tag{7.4-26}$$

We can then solve for z_1, z_2, and z_R in terms of the mirror curvatures (R_1 and R_2) and the spacing between the mirrors (d). The solutions are

$$z_1 = -\frac{d(1 - g_1)g_2}{g_1 + g_2 - 2g_1 g_2}, \tag{7.4-27}$$

$$z_2 = \frac{d(1 - g_2)g_1}{g_1 + g_2 - 2g_1 g_2}, \tag{7.4-28}$$

$$z_R^2 = \frac{d^2 g_1 g_2 (1 - g_1 g_2)}{(g_1 + g_2 - 2g_1 g_2)^2}, \tag{7.4-29}$$

where $g_1 = 1 + d/R_1$ and $g_2 = 1 + d/R_2$ are defined in Chapter 2 [after Eq. (2.4-42)].

The waist and the beam widths at the two mirrors are then

$$w_0^2 = \frac{2d}{k_0} \sqrt{\frac{g_1 g_2(1 - g_1 g_2)}{(g_1 + g_2 - 2g_1 g_2)^2}} , \qquad (7.4\text{-}30)$$

$$w^2(z = z_1) = w_1^2 = \frac{2d}{k_0} \sqrt{\frac{g_1}{g_2(g_1 - g_1 g_2)}} , \qquad (7.4\text{-}31)$$

$$w^2(z = z_2) = w_2^2 = \frac{2d}{k_0} \sqrt{\frac{g_1}{g_2(g_1 - g_1 g_2)}} . \qquad (7.4\text{-}32)$$

Note that real and finite solutions for the Gaussian beam parameters above can exist only if the g parameters satisfy the stability condition, $0 \le g_1 g_2 \le 1$, which has been derived [see Eq. (2.4-42)] using the geometrical optics approach.

We conclude with the observation that in a practical laser, one of the mirrors has a nonzero transmission coefficient. Thus, the transmission characteristics will depend on the mode number as well as the finesse of the cavity (see Figure 3.8). Furthermore, the resonances will be amplitude-modulated by the line-shape function.

7.5 The Helium – Neon Laser

The helium–neon (He–Ne) laser is one of the most popular and most widely used gas lasers. It was the first gas laser fabricated. The He–Ne laser consists of an approximately $10:1$ mixture of helium and neon placed inside a long narrow discharge tube at a pressure of about 1 torr ($\sim$ 1 mm of mercury). This gas mixture is the *lasing medium* that provides population inversion. The gas system is enclosed between a set of mirrors so that a *resonator* can be formed. *Pumping* is achieved by an electrical discharge produced by a high voltage ($\sim$ 1–2 kV). The three essential elements of the He–Ne laser, that is, the pump, lasing medium, and resonator, are shown schematically in Figure 7.9.

Figure 7.10 shows the energy-level transition of a He–Ne laser. The actual lasing action occurs between energy levels of neon, whereas the helium is included to assist the pumping process. There are two major types of excitation in the pumping process:

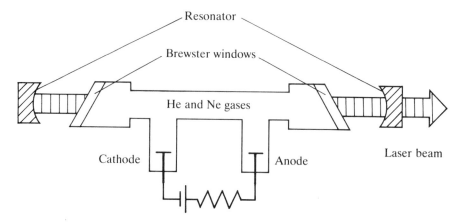

Figure 7.9 Typical He–Ne system.

direct electron and atomic collision excitation. In the former, free electrons are accelerated down the tube through a discharge. These electrons collide with the helium atoms and excite them to two metastable energy levels F_2 and F_3. This is represented by

$$e_1 + \text{He} = \text{He}^* + e_2 \quad \text{(electron collision)}, \qquad (7.5\text{-}1)$$

where e_1 and e_2 are the electron energies before and after the collision, and the asterisk denotes an excited atom. Because some of the excited states of neon, namely, $\tilde{E}_5$ and $\tilde{E}_3$, correspond approximately to the same energy as that of the excited levels F_3 and F_2 of helium, the helium atoms can then transfer their energy to neon atoms through atomic collisions, thereby exciting the neon atoms in the ground level $\tilde{E}_0$ to levels $\tilde{E}_3$ and $\tilde{E}_5$. This, of course, brings the helium atoms back to the ground level. The energy transfer is represented by

$$\text{He}^* + \text{Ne} = \text{Ne}^* + \text{He} \quad \text{(atomic collision)}, \qquad (7.5\text{-}2)$$

where the excited atoms again are denoted by an asterisk. Because levels F_2 and F_3 are metastable, this process of energy transfer has a high probability. Therefore, the discharge through the gas mixture continuously populates energy levels $\tilde{E}_3$ and $\tilde{E}_5$, thereby helping to create population inversion between levels $\tilde{E}_3$ and $\tilde{E}_5$, and levels

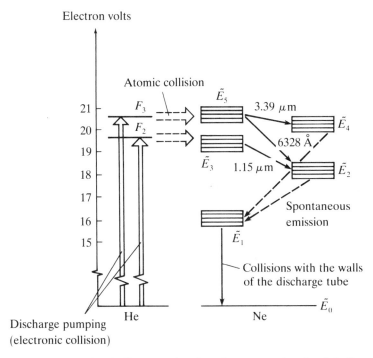

Figure 7.10 Energy level diagram showing the energy levels of helium and neon.

$\tilde{E}_4$ and $\tilde{E}_2$, as shown in Figure 7.10. Transitions between levels $\tilde{E}_5$ and $\tilde{E}_3$ and between levels $\tilde{E}_4$ and $\tilde{E}_2$ are forbidden by quantum mechanical selection rules, and transitions from levels $\tilde{E}_4$ and $\tilde{E}_2$ to $\tilde{E}_1$ are spontaneous so that the population inversion is maintained. The level $\tilde{E}_1$ is a metastable state, but a buildup of atoms in this state is not possible due to collisions with the walls of the discharge tube, which lowers their energy level to $\tilde{E}_0$. From the preceding discussion, we can immediately recognize that He–Ne lasers are four-level systems with $\tilde{E}_3$ and $\tilde{E}_5$ the upper laser levels and $\tilde{E}_2$ and $\tilde{E}_4$ the lower laser levels. The transitions from $\tilde{E}_5$ to $\tilde{E}_4$, $\tilde{E}_3$ to $\tilde{E}_2$, and $\tilde{E}_5$ to $\tilde{E}_2$ lead to emissions at wavelengths of 3.39 μm, 1.15 μm, and 6328 Å, respectively. Specific frequency selection can be made by choosing multilayer dielectric mirrors giving a maximum reflectivity at the desired wavelength. If polarization of the output laser light is required, the discharge tube can be provided with

Brewster-angle end windows, as shown in Figure 7.9. Typical output power of He–Ne lasers ranges from 1–50 mW with input electrical power of about 5–10 W.

7.6 Photodetectors

So far in this chapter, we have examined sources of coherent light. In what follows, we will study the behavior of devices that convert optical radiation to electrical signals. We will look at two distinct photodetection mechanisms: the *external photoelectric effect* and the *internal photoelectric effect*. In the former, electrons are emitted from the surface of a metal when illuminated by photons having sufficient energies. The vacuum photodiode and the photomultiplier are based on this effect. In the internal photoelectric effect, free charge carriers (i.e., electrons and holes) are generated around the semiconductor junction by absorption of incoming photons. The p–n junction diode, PIN diode, and avalanche photodiode (APD) are the three common devices using this phenomenon.

7.6.1 The Vacuum Photodiode and Photomultiplier

Figure 7-11a shows a schematic diagram of a vacuum photodiode. A photoemissive surface (commonly called the *photocathode*) is placed inside a vacuum tube with another electrode (the *anode*). In order to be freed from the cathode surface, the (photo) electron must acquire a minimum amount of energy called the *work function* Φ. The work function is the minimum energy required to free an electron from the conduction band of a metal and bring it to the vacuum level. Therefore, an incident photon must possess at least this amount of energy to cause photoemission. Quantitatively, we have

$$\hbar\omega_0 \geq \Phi, \qquad (7.6\text{-}1)$$

which conveniently gives the *cutoff wavelength*

$$\lambda_c = \frac{1.24}{\Phi}, \qquad (7.6\text{-}2)$$

where λ_c is measured in microns and Φ is given in electron volts (eV). Incident wavelengths longer than λ_c cannot be detected.

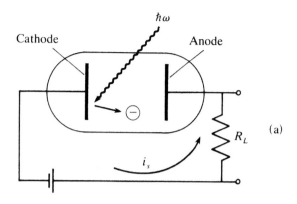

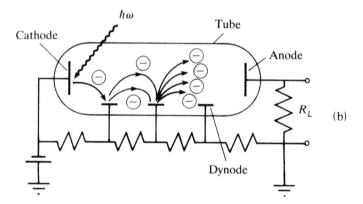

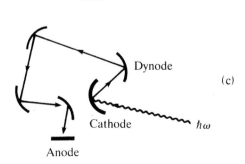

Figure 7.11 (a) Vacuum photodiode. (b) Photomultiplier. (c) Photomultiplier with focusing mechanism.

However, not every photon whose energy is greater than the work function will free an electron. The ratio of the number of emitted electrons to the number of absorbed photons is called the *quantum efficiency* η_q. Therefore, the maximum quantum efficiency, by definition, is unity. Pure metals are rarely used as practical photocathodes as they have low quantum efficiency and high values for the work function. Cesium is a common photoemissive material, having a work function of 1.9 eV. The photocurrent i_s generated by incident light of power P_i is given by

$$i_s = \frac{\eta_q e P_i}{\hbar \omega_0}, \qquad (7.6\text{-}3)$$

where the term $P_i/\hbar\omega_0$ is the rate at which photons strike the photocathode, $\eta_q P_i/\hbar\omega_0$ is the number of emitted electrons per second, and e is the electron charge (1.6×10^{-19} C). Hence, the vacuum diode acts as if it were a current source for the receiving circuit. The *responsivity* $\mathcal{R}$ of the photodetector (measured in amperes per watt) is defined by

$$\mathcal{R} \triangleq \frac{i_s}{P_i} = \frac{\eta_q e}{\hbar \omega_0} = 8.1 \times 10^5 \lambda_{\nu} \eta_q \quad \left(\frac{A}{W} \right), \qquad (7.6\text{-}4)$$

which yields a typical responsivity of about 6.4 mA/W at $\lambda_{\nu} = 0.8$ μm with $\eta_q = 1\%$. The output voltage V_0 measured across a load resistance of R_L is

$$V_0 = i_s R_L = \mathcal{R} P_i R_L. \qquad (7.6\text{-}5)$$

Equations (7.6-3)–(7.6-5) are valid for the external photoelectric effect as well as for the internal photoelectric effect, and we note that, from Eq. (7.6-3), the photocurrent output is proportional to the power incident on the surface of the photodetector.

Generally, the output of a simple vacuum diode is relatively small and requires amplification, which is achieved by employing a phototube filled with a gas such as argon under low pressure. A photoelectron accelerated by the anode collides with the gas atoms and hence ionizes them, thereby generating more electrons. The overall current gain is typically of the order of 10. Internal gain also can be achieved in a photomultiplier, where the photoelectrons are accelerated towards the anode through a series of electrodes (called *dynodes*). The interstage voltage gradients for the dynodes are supplied by a voltage-divider network consisting of a series of connected resistors, as shown in Figure 7.11(b). When accelerated photoelectrons emitted from the photocathode hit the dynode, they release electrons. This process is called *secondary emission* and repeats itself at each dynode. If m is the average number of secondary electrons emitted at each dynode surface for each incident electron, and if there are N dynodes in the system, the total gain is then

$$G = m^N \qquad (7.6\text{-}6)$$

and the current through the external circuit is

$$i_s = \frac{G\eta_q e P_i}{\hbar\omega_0}. \qquad (7.6\text{-}7)$$

For typical values of $N = 9$ and $m = 5$, current amplification is almost 2×10^6. Internal amplification has an important advantage as it increases the signal level without significantly decreasing the signal-to-noise power ratio, whereas external amplification, such as the use of electronic amplifiers, always adds noise to the system.

The speed of response is determined by the time for the electron to traverse the dynode chain from cathode to anode (*electron transit time*). Because different electrons take different trajectories through the tube, this introduces a spread in the transit time, which gives rise to the so-called *rise time* and limits the ability of the photomultiplier to respond faithfully to a short optical pulsed signal. Transit-time spread may be reduced by employing fewer dynodes and carefully designing the dynodes (such as the focussing dynode configuration shown in Figure 7.11(c)) so that the electrons can take similar paths. Some photomultipliers have rise times of a few tenths of a nanosecond and transit times of about 30 ns. However, even if the photomultiplier tube has a fast response time, the response will be limited to the 3-dB bandwidth $\omega_{3\text{-dB}} = (R_L C_S)^{-1}$ of the output circuit, where C_S denotes the total capacitance between the tube anode and all other electrodes, including stray capacitances such as wiring capacitances.

7.6.2 Semiconductor Junction Detectors

We will describe three types of junction detectors: the p–n, PIN, and avalanche photodiodes. The simple p–n diode is illustrated in Figure 7.12(a). The diode junction is reverse-biased, causing mobile holes and electrons to move away from the junction. Diffusion of holes (electrons) from the p (n) region toward the n (p) region leaves behind the immobile negative acceptor ions (positive donor ions), which, in turn, establish an electric field distribution in the vicinity of the junction called the *depletion region*, as shown in Figure 7.12(b). Because there are no free charges, the resistance in

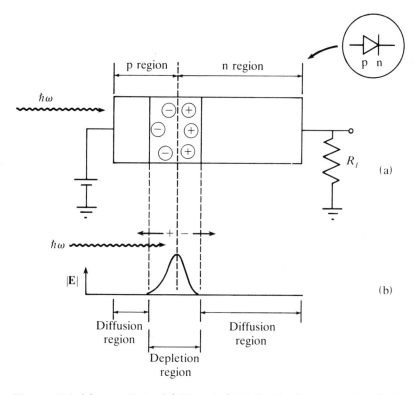

Figure 7.12 (a) p–n diode. (b) Electric field distribution across the diode.

this region is high, so that the voltage drop across the diode mostly occurs across the junction. When an incident photon is absorbed in the depletion region after passing through the p layer, it raises an electron from the valance to the conduction band. The electron is now free to move, and a hole is left in the valance band. In this way, free charge carrier pairs, commonly called *photocarriers*, are created by photon absorption. These moving carriers then cause current flow through the external circuit. In order to generate the electron–hole pair, the incident photon must have energy larger than that of the band gap $\tilde{E}_g$ between the valence and the conduction bands, that is, $\hbar\omega_0 \geq \tilde{E}_g$. In terms of the cutoff wavelength λ_c, we have

$$\lambda_c = \frac{1.24}{\tilde{E}_g}, \tag{7.6-8}$$

with $\tilde{E}_g$ measured in electron volts. This equation is like Eq. (7.6-2) for the external photoelectric effect. The cutoff wavelength is about 1.06 μm for silicon and 1.6 μm for germanium, where their band-gap energies are 1.1 and 0.67 eV, respectively. Just as in the case of the external photoelectric effect, not all wavelengths lower than λ_c can generate photocarriers, as the absorption of photons in the p and n regions is increased at the shorter wavelengths. After photocarriers are generated in the p or n region, most of the free carriers will, in effect, diffuse randomly through the diode and recombine before reaching the depletion junction. The quantum efficiency η_q for the semiconductor junction diode can then be defined, similarly to that in the photoemissive situation, as the number of electron–hole pairs generated per incident photon [see Eq. (7.6-3)].

Two main factors limit the *response time* of a photodiode:

1. *the transit time of the photocarriers through the depletion region;*
2. *the diffusion time of the photocarriers (generated in the depletion region) through the diffusion region.*

Carrier diffusion is inherently a slow process. In order to have a high-speed photodiode, the carriers should be generated in the depletion region in the high field intensity area or close to it so

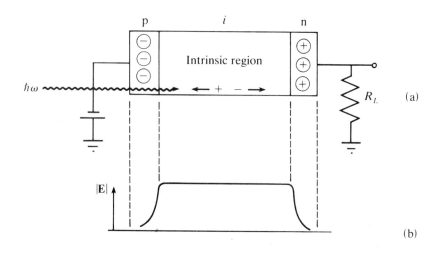

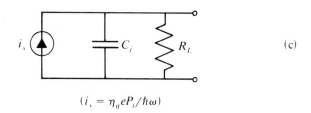

$$(i_s = \eta_q e P_i / \hbar \omega)$$

Figure 7.13 (a) PIN diode. (b) Electric field distribution. (c) Equivalent circuit of a PIN diode.

that the diffusion times are less than the carrier transit times. This can be accomplished by increasing the bias voltage, but practical constraints limit the applied bias voltage. Another method is to add an intrinsic region between the p and n regions, as shown in Figure 7.13(a), where we have a popular PIN diode structure. Figure 7.13(b) shows its corresponding electric field intensity $|E|$ as a function of distance. Note that in the intrinsic region the field intensity is high. When a photon is injected into the intrinsic region, an electron–hole pair is created. The hole drifts towards the p region and the electron towards the n region, thereby generating a current flow in the external circuit. The use of a wide intrinsic layer improves the quantum efficiency because the incident photons will be absorbed in it rather than in the thin depletion region as in the p–n diode. Also, it lowers the junction capacitance $C_j = \epsilon_i A / W_i$, where ϵ_i and W_i denote the permittivity and the width of the intrinsic layer, respectively, and A is the area of the junction.

Just as the speed of response in p–n diodes is limited by the transit time of the photocarriers through the depletion region, the speed of response in PIN diodes is limited by the transit time of the carriers through the intrinsic region. Increasing the bias voltage can reduce the transit time as the velocity of the free charge carriers is proportional to the magnitude of the bias voltage. Junction capacitance also limits the response. An equivalent circuit of a PIN diode with its associated junction capacitance is shown in Figure 7.13(c). The 3-dB bandwidth is

$$\omega_{3\text{-dB}} = \frac{1}{R_L C_j}, \qquad (7.6\text{-}9)$$

where we have neglected stray capacitances such as the capacitance of the packaging structure. The corresponding *circuit rise time* is then approximately given by

$$t_r = \frac{2\pi}{\omega_{3\text{-dB}}} = 2\pi R_L C_j. \qquad (7.6\text{-}10)$$

The speed of response of a photodiode together with its output circuit [see Figure 7.13(c)] may be limited by the carrier transit time or the circuit rise time, whichever is larger.

To achieve internal gain for junction diodes, that is, to devise a means for generating more than one electron–hole pair for each detected photon, one uses the *avalanche effect*. Figure 7.14(a) shows the structure of an *avalanche photodiode* (APD). The p^+ and n^+ layers are highly doped. The p^- region is lightly doped and nearly intrinsic. When photons are absorbed in this region, electron–hole pairs are created. As in the PIN diode, the holes and electrons drift towards the p^+ and n^+ regions, respectively. How-

Lasers and Photodetectors

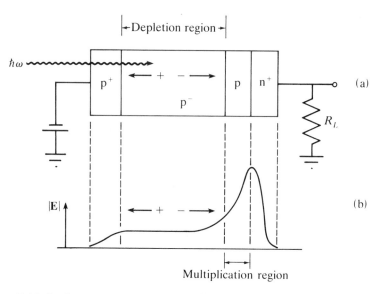

Figure 7.14 Avalanche photodiode: (a) junction configuration; (b) electric field distribution.

ever, before the electron reaches the n^+ region, it passes through a p-type multiplying region. In this region the electron is accelerated by the high $|\mathbf{E}|$ field as shown in Figure 7.14(b). If the electron undergoes sufficient acceleration in the region, new electron–hole pairs are created by the collision–ionization process. These newly created electron–hole pairs themselves generate more electron–hole pairs, thus creating the so-called *avalanche multiplication*. In this way, the number of electron–hole pairs created may be multiplied by a factor of 1000 over the traditional PIN photodiodes. Therefore, APD diodes have an internal gain for the induced photocurrent and are dominantly used in high-speed systems because of their high sensitivity. Note that APDs require a circuit for the high reverse-bias voltage that ranges from tens to hundreds of volts, whereas in PIN diode circuits a few volts of reverse-bias are usually sufficient. Hence, the PIN diode is preferable in most laboratory systems. However, the APD gain is needed when the system is loss-limited, which may occur for long-distance fiber optic links.

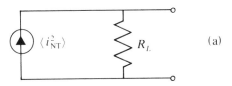

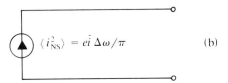

Figure 7.15 (a) Thermal-noise equivalent circuit. (b) Shot-noise equivalent circuit.

7.7 Noise Mechanisms

In this section we classify different sources of noise during photodetection and calculate the signal-to-noise ratio. We consider two major classes of noise: *thermal noise* and *shot noise*.

Thermal noise is also called *Johnson noise* or *Nyquist noise*. It originates within the load resistor R_L in the photodetector circuit and arises because of the thermal energy of electrons within the resistor. The random nature of the electron motion results in a fluctuating current through the resistor. The time-averaged value of the thermal-noise current $\langle i_{NT} \rangle$ is zero, but the mean-square value of this current $\langle i_{NT}^2 \rangle$ having frequency components between ω and $\omega + \Delta\omega$ at a temperature T (measured in kelvins) is given by

$$\langle i_{NT}^2 \rangle = \frac{2kT\,\Delta\omega}{\pi R_L}, \qquad (7.7\text{-}1)$$

where the brackets $\langle\ \rangle$ indicate a time average. The thermal-noise equivalent circuit of a resistor is shown in Figure 7.15(a), where R_L is now an ideal noiseless resistor. The preceding equation has been derived under the assumption that the thermal-noise spectrum is

Lasers and Photodetectors

uniform. The average noise power $\bar{P}_{NT}$ generated within the resistor is then

$$\bar{P}_{NT} = \langle i_{NT}^2 \rangle R_L = \frac{2kT\,\Delta\omega}{\pi}. \qquad (7.7\text{-}2)$$

Shot noise is also called *quantum noise* and arises directly from the discrete nature of electrons and from the random generation and collection of photoelectrons (from a photoemissive surface) or random generation and recombination of photocarriers (in a semiconductor junction). The fluctuations follow Poisson statistics and, because this is a fundamental property of the photodetection process, the presence of shot noise sets a lower limit on the minimum detectable signal. We can show that the mean-square value of the shot noise with frequencies between ω and $\omega + \Delta\omega$ is given by

$$\langle i_{NS}^2 \rangle = \frac{e\bar{i}\,\Delta\omega}{\pi}, \qquad (7.7\text{-}3)$$

where $\bar{i}$ is the average current flowing in the photodetector and is the sum of the average signal current $\langle i_s \rangle$ due to the optical signal power and the *dark current* i_d. The dark current may be due to the thermionic emission from the cathode of a vacuum photodiode or the thermal generation of electron–hole pairs in a semiconductor. Shot noise can be represented by an equivalent circuit consisting of an ideal current source, as shown in Figure 7-15(b). Note that the shot noise current increases with $\langle i_s \rangle$ and, therefore, with an increase in the incident optical power. This differs from the thermal noise, which is independent of the optical power.

With the preceding equivalent circuits of the thermal-noise and shot-noise sources, we can now evaluate the performance of photodetector circuits in terms of the signal-to-noise ratio. The photodetector receiving circuit, including the equivalent sources of noise, is shown in Figure 7.16. We assume that the incident optical power P_i is constant. Therefore, the signal photocurrent is constant, given by

$$i_s = \frac{\eta_q e P_i}{\hbar \omega}. \qquad (7.7\text{-}4)$$

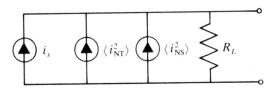

Figure 7.16 Equivalent circuit of a photodetector, including the sources of thermal and shot noise.

The photodetector then delivers constant electrical signal power

$$\bar{P}_E = i_s^2 R_L = \left(\frac{\eta_q e P_i}{\hbar\omega}\right)^2 R_L \qquad (7.7\text{-}5)$$

to the load resistor. The average thermal-noise power delivered to the load is given by Eq. (7.7-2), and the shot-noise power is

$$\bar{P}_{NS} = \langle i_{NS}^2 \rangle R_L = \frac{e\bar{i}\,\Delta\omega\,R_L}{\pi}, \qquad (7.7\text{-}6)$$

where

$$\bar{i} = \langle i_s \rangle + i_d, \qquad (7.7\text{-}7)$$

and where $\langle i_s \rangle = i_s$ is given by Eq. (7.7-4).

Now, defining the *signal-to-noise power ratio* S/N as the average signal power divided by the average power due to all noise sources, we have

$$\frac{S}{N} = \frac{\bar{P}_E}{\bar{P}_{NT} + \bar{P}_{NS}}$$

$$= \frac{\left(\eta_q e P_i/\hbar\omega\right)^2 R_L}{2kT\,\Delta\omega/\pi + 2e\left[\left(\eta_q e P_i/\hbar\omega\right) + i_d\right]R_L}. \qquad (7.7\text{-}8)$$

Let us consider two special cases. Assuming that the average signal power is much larger than the dark current (i.e., $\eta_q e P_i/\hbar\omega \gg i_d$) and that the shot-noise power far exceeds the

Lasers and Photodetectors

thermal-noise power, Eq. (7.7-8) simplifies to

$$\frac{S}{N} = \frac{\pi \eta_q P_i}{\hbar \omega \, \Delta \omega} . \tag{7.7-9}$$

This S/N is called *shot-noise limited* and is the best result obtainable. Defining that the *minimum detectable power* $(P_i)_{min}$ is given by S/N = 1, we have

$$(P_i)_{min} = \frac{\hbar \omega \, \Delta \omega}{\pi \eta_q} . \tag{7.7-10}$$

One can obtain the shot-noise limited S/N by increasing the incident optical power, thereby eliminating the effects of dark current and thermal noise. However, in most practical cases, we do not always have unlimited power. When the optical power is low, thermal noise usually dominates over shot noise. Equation (7.7-8) then reduces to

$$\frac{S}{N} = \frac{\pi (\eta_q e P_i / \hbar \omega)^2 R_L}{2kT \, \Delta \omega} . \tag{7.7-11}$$

This S/N is *thermal-noise limited* and is usually much smaller than the shot-noise limited S/N. Note that this limit can be improved by employing a larger load resistor or by lowering the temperature of the photodetector system. Note also that both the shot-noise limited and thermal-noise limited S/N can be improved by narrowing the bandwidth·of the photodetector system.

7.8 Detection Schemes

In this section we describe two detection schemes and compare their S/N ratios. The two schemes are *direct detection* and *heterodyne detection*.

Figure 7.17 shows a direct-detection optical receiver. The input signal is given by

$$s(t) = A_0 \cos(\omega_0 t + \theta_0). \tag{7.8-1}$$

Because the photodetector responds to the power of the optical

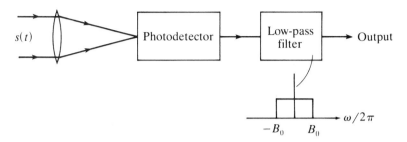

Figure 7.17 Direct detection scheme.

signal only, the average photocurrent of the photodetector is

$$\langle i_s \rangle = \mathscr{R} P_i,\qquad(7.8\text{-}2)$$

where $\mathscr{R}$ is the responsivity of the photodetector defined as the detector power-to-current conversion factor. For detectors with no internal gain, it is given by Eq. (7.6-4). P_i is the average incident power defined by

$$P_i = \langle s^2(t) \rangle_\tau = \tfrac{1}{2} A_0^2,\qquad(7.8\text{-}3)$$

where the brackets and the subscript τ indicate a time average taken over the photodetector response time τ. Hence, the shot-noise limited S/N ratio for the direct detection, according to Eqs. (7.7-9) and (7.8-3), is

$$\frac{S}{N} = \frac{\pi \eta_q}{\hbar \omega_0 \, \Delta \omega} P_i = \frac{\eta_q A_0^2}{4 \hbar \omega_0 B_0},\qquad(7.8\text{-}4)$$

where $\Delta \omega = 2\pi B_0$ is due to the low-pass filter at the output stage.

In heterodyne detection, the input signal is mixed with a local oscillator (LO) reference $r(t)$ through the half-silvered mirror, as shown in Figure 7.18. The LO reference signal is

$$r(t) = A_r \cos(\omega_r t + \theta_r)\qquad(7.8\text{-}5)$$

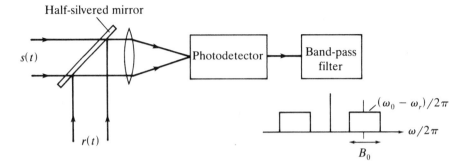

Figure 7.18 Heterodyne detection scheme.

and the average power incident on the photodetector is given by

$$P_i = \langle |s(t) + r(t)|^2 \rangle_\tau$$

$$= \frac{A_0^2}{2} + \frac{A_r^2}{2} + 2A_0 A_r \langle \cos(\omega_0 t + \theta_0)\cos(\omega_r t + \theta_r) \rangle_\tau. \quad (7.8\text{-}6)$$

Now

$$\langle \cos(\omega_0 t + \theta_s)\cos(\omega_r t + \theta_r) \rangle_\tau = \tfrac{1}{2}\langle \cos[(\omega_0 + \omega_r)t + (\phi_0 + \phi_r)]$$

$$+ \cos[(\omega_0 - \omega_r)t + (\theta_0 - \theta_r)] \rangle_\tau.$$

$$(7.8\text{-}7)$$

The first term is time-averaged to zero as τ is much larger than the period of the signal being averaged, that is, $\tau \gg 2\pi(\omega_0 + \omega_r)^{-1}$. For the second term, if we assume that $\omega_0 \approx \omega_r$, we have $\tau \ll 2\pi(\omega_0 - \omega_r)^{-1}$ and the second term within the time average can be considered as a constant with respect to the time-averaging. Therefore, Eq. (7.8-6) is reduced to

$$P_i = \frac{A_0^2}{2} + \frac{A_r^2}{2} + A_0 A_r \cos[(\omega_0 - \omega_r)t + \phi_0 - \phi_r]. \quad (7.8\text{-}8)$$

The photocurrent is then

$$i_s = \mathscr{R}P_i$$

$$= \mathscr{R}\left[\frac{A_0^2}{2} + \frac{A_r^2}{2} + A_0 A_r \cos\left[(\omega_0 - \omega_r)t + \phi_0 - \phi_r\right]\right]. \quad (7.8\text{-}9)$$

The first two terms in Eq. (7.8-9) are dc contributions, whereas the last term is a heterodyne current at frequency $\omega_0 - \omega_r$. The average electric power delivered to the load is then

$$\bar{P}_E = \langle i_s^2 \rangle R_L = \tfrac{1}{2}\mathscr{R}^2 A_0^2 A_r^2 R_L. \quad (7.8\text{-}10)$$

In order to calculate the shot-noise limited S/N, we first find $\bar{P}_{NS}$. From Eqs. (7.7-6) and (7.7-7), and assuming $\langle i_s \rangle \gg i_d$, we have

$$\bar{P}_{NS} = 2e\langle i_s \rangle B_0 R_L = e\mathscr{R}\left[A_0^2 + A_r^2\right]B_0 R_L, \quad (7.8\text{-}11)$$

where we have used Eq. (7.8-9) to find $\langle i_s \rangle$.

The S/N is then calculated from Eq. (7.7-8), and we have

$$\frac{S}{N} = \frac{\tfrac{1}{2}\mathscr{R}^2 A_0^2 A_r^2 R_L}{e\mathscr{R}\left[A_0^2 + A_r^2\right]B_0 R_L + 4kTB_0} \quad (7.8\text{-}12)$$

Employing a strong LO signal power (i.e., $A_r^2 \gg A_0^2$), shot noise becomes dominant and Eq. (7.8-12) simplifies to

$$\frac{S}{N} = \frac{\eta_q A_0^2}{2\hbar\omega B_0} \quad (7.8\text{-}13)$$

We note that the shot-noise limited S/N in the heterodyne detection case is a factor of 2 larger than that in the direct detection scheme [see Eq. (7.8-4)]. We conclude that optical heterodyne receivers can be very sensitive. They approach the ideal S/N even at low signal levels.

Problems

7.1 If $B_{12} \neq B_{21}$, that is, in the presence of degeneracies, find $\int \alpha(\omega) \, d\omega$ in terms of $\tilde{g}_1$ and $\tilde{g}_2$, where $\tilde{g}_1$ and $\tilde{g}_2$ are the degeneracies for levels 1 and 2.

7.2 Plot the dimensionless function $g(\omega) \, \Delta\omega$ versus dimensionless frequency offset $2(\omega - \omega_0)/\Delta\omega$ for the normalized Lorentzian and Gaussian line shapes, assuming that $\Delta\omega_h = \Delta\omega_D$. Can you make some remarks comparing the two plots?

7.3 For any two broadening processes that are uncorrelated with each other, that is, they are simultaneously present but independent, it can be shown that the overall line-shape function is given by the convolution between the two processes.
(a) Show that the convolution of two Gaussian line shapes of widths $\Delta\omega_1$ and $\Delta\omega_2$ gives a Guassian line shape of width $\Delta\omega = \sqrt{\Delta\omega_1^2 + \Delta\omega_2^2}$.
(b) Show that the convolution of a Lorentzian line shape of width $\Delta\omega_1$ with another Lorentzian line shape of width $\Delta\omega_2$ gives a Lorentzian of width $\Delta\omega = \Delta\omega_1 + \Delta\omega_2$.

7.4 Set up an integral equation similar to Eq. (7.4-2) for a resonator with two flat mirrors. Assume that the separation between the mirrors is d.

7.5 The field distribution along the plane AB in Figure 7.6 is given by Eq. (7.4-15). Verify that the field in the plane midway between the mirrors is given by Eq. (7.4-22).

7.6 Sketch the mode patterns for the following cases:
(a) TEM_{20},
(b) TEM_{12},
(c) TEM_{22},
(d) TEM_{03},
(e) TEM_{13}.

7.7 Determine the responsivity of a photodetector having a quantum efficiency of 1% at $\lambda_t = 0.8 \ \mu m$. If a light beam of

1 μW optical power is incident on the photodetector, calculate the voltage across a 50-Ω load resistor.

7.8 Find the S/N of a photodetector with internal gain M and show that if the gain is large enough, the ideal shot-noise limited S/N is given by Eq. (7.7-9).

7.9 The power in an amplitude modulated (AM) light beam is given by

$$P(t) = P_0(1 + m \cos \omega_m t),$$

where m is the modulation index and ω_m is the modulation frequency. The beam of frequency ω is incident on a PIN photodiode with quantum efficiency η_q.
(a) Find the rate at which photons hit the diode and the rate at which carriers are generated.
(b) Find an expression for the signal current and calculate the average electrical signal power delivered to the load resistor R_L.
(c) Find the thermal-noise limited and the shot-noise limited S/N ratios.

7.10 Repeat Problem 7.9 if the diode is replaced by a photomultiplier with gain M and quantum efficiency η_q.

7.11 Defining the theoretical minimum detectable optical power P_{min} by setting S/N = 1, calculate P_{min} for the direct and heterodyne detection under the following conditions:

optical frequency $\nu = 6 \times 10^{14}$ Hz;

quantum efficiency $\eta = 10\%$;

filter bandwidth $\Delta\nu = 1$ Hz.

References
7.1 Fowles, G. R. (1975). *Introduction to Modern Optics*. Holt, Rinehart and Winston, New York.
7.2 Ghatak, A. (1977). *Optics*, TATA McGraw-Hill, New Delhi.
7.3 Ghatak, A. and K. Thyagarajan (1978). *Contemporary Optics*. Plenum, New York.

7.4 Ghatak, A. and K. Thyagarajan (1981). *Lasers*: *Theory and Applications*. Plenum, New York.

7.5 Iizuka, K. (1985). *Engineering Optics*. Springer-Verlag, Berlin.

7.6 Keiser, G. (1983). *Optical Fiber Communications*. McGraw-Hill, New York.

7.7 Kittel, C. (1976). *Introduction to Solid State Physics*, 5th Ed. John Wiley and Sons, New York.

7.8 Lengyel, B. A. (1962). *Lasers*: *Generation of Light by Stimulated Emission*. John Wiley and Sons, New York.

7.9 Messiah, A. (1961). *Quantum Mechanics*. Interscience, New York.

7.10 Milonni, P. W. and J. H. Eberly (1988). *Lasers*. John Wiley and Sons, New York.

7.11 Palais, J. C. (1988). *Fiber Optics Communications*, 2 Ed. Prentice-Hall, Englewood Cliffs, New Jersey.

7.12 Siegman, A. E. (1986). *Lasers*. University Science Books, Mill Valley, California.

7.13 Slepian, D. and H. O. Pollak (1961). Prolate spheroidal wave functions, Fourier analysis and uncertainty—I. *Bell Syst. Tech. J.* **40** 43.

7.14 Yariv, A. (1976). *Introduction to Optical Electronics*. 2d Ed. Holt, Rinehart, and Winston, New York.

Chapter 8 Nonlinear Optics

Thus far, all our discussions on optical propagation and optical processing were limited to linear media. In this chapter, we take a simplified look at the vast area of nonlinear optics, which deals with optical propagation through nonlinear media. We first introduce the concept of nonlinearity in the simplest possible way. We than discuss some of the effects of nonlinearity, namely, second-harmonic (and subharmonic) generation, self-refraction, bistability, phase conjugation, and soliton propagation in optical fibers.

8.1 The Concept of Nonlinearity

In this section, we will first expose readers to a very simple model of wave propagation in a nonlinear medium. Thereafter, we will relate the nonlinear parameters that we have introduced to the commonly used parameters in nonlinear optics. A short discussion on the physical origin of these nonlinearities will then be presented.

Let us begin by postulating that in a nonlinear medium, the *nonlinearity* can be attributed to the *dependence of the phase velocity on the amplitude of the propagating wave*, unlike the linear case. Recall, from Chapter 3, that a PDE, for a wavefunction $\psi(z, t)$, of the form

$$\frac{\partial^2 \psi}{\partial t^2} - v^2 \frac{\partial^2 \psi}{\partial z^2} = 0 \qquad (8.1\text{-}1a)$$

Nonlinear Optics

models bidirectional wave propagation in a linear nondispersive medium [see Eqs. (3.2-18) and (3.2-19)]. However, unidirectional propagation (viz., along $+z$) can be effectively modeled by

$$\frac{\partial \psi}{\partial t} + v \frac{\partial \psi}{\partial z} = 0, \qquad v > 0. \tag{8.1-1b}$$

In a nonlinear medium, the phase velocity v is modified to v' according to

$$v' = v\left(1 + \beta_2 \psi + \beta_3 \psi^2 + \cdots\right) \tag{8.1-2}$$

where β_2 and β_3 are constants [Whitham (1974)]. Thus, in the nonlinear regime, Eq. (8.1-1b) is modified, in light of Eq. (8.1-2), to

$$\frac{\partial \psi}{\partial t} + v\left(1 + \beta_2 \psi + \beta_3 \psi^2\right) \frac{\partial \psi}{\partial z} = 0 \tag{8.1-3a}$$

or

$$\frac{\partial \psi}{\partial t} + v \frac{\partial \psi}{\partial z} + \left(\frac{v \beta_2}{2}\right) \frac{\partial \psi^2}{\partial z} + \left(\frac{v \beta_3}{3}\right) \frac{\partial \psi^3}{\partial z} = 0. \tag{8.1-3b}$$

Note that Eq. (8.1-3b) is a *nonlinear* PDE! The constants β_2 and β_3 are called the *quadratic* and *cubic nonlinearity coefficients* because they appear with the terms $\partial \psi^2/\partial z$ and $\partial \psi^3/\partial z$, respectively. The analysis of nonlinear PDEs is what makes the subject of nonlinear optics more challenging than linear optics, which we have studied thus far.

In fluid mechanics, the nonlinearity (e.g., β_2) is responsible for *shock formation* during fluid flow due to steepening of the (baseband) wave profile as shown in Figure 8.1. This happens because the parts of the baseband pulse having larger amplitudes travel faster than the parts with smaller amplitudes for $\beta_2 > 0$, leading to a point in time where the right edge of the pulse develops infinite steepness, or *shock*. In optics, the quadratic non-linearity is responsible for second-harmonic generation, and this

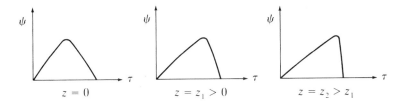

Figure 8.1 Evolution of shock during propagation of a pulse, assuming $\beta_2 > 0$; $\tau = t - z/v$ is a moving frame of reference.

will be discussed in Section 8.2. Special effects arising from the cubic nonlinearity (e.g., self-refraction, bistability, phase conjugation, and soliton propagation) will be discussed at length in Sections 8.3–8.7.

The simple model described previously cannot, however, describe wave propagation in higher dimensions (x, y, z), necessitating a higher-order PDE. As an illustrative example, we demonstrate how to derive such a PDE for $\beta_2 \neq 0$, $\beta_3 = 0$. First, we differentiate Eq. (8.1-3b) with respect to t to get [Korpel and Banerjee (1984)]

$$\frac{\partial^2 \psi}{\partial t^2} + v \frac{\partial}{\partial z}\left[\frac{\partial \psi}{\partial t}\right] + v\beta_2\left[\psi \frac{\partial \psi}{\partial t}\right] = 0. \qquad (8.1\text{-}4)$$

Now, we substitute for $\partial \psi / \partial t$ from Eq. (8.1-3a) and neglect higher-order terms in β_2 (assuming weak nonlinearity) to obtain

$$\frac{\partial^2 \psi}{\partial t^2} - v^2 \frac{\partial^2 \psi}{\partial z^2} \approx v^2 \beta_2 \frac{\partial^2 \psi^2}{\partial z^2}. \qquad (8.1\text{-}5)$$

Because the nonlinearity has been assumed to be weak, we can replace z by vt on the right-hand side of Eq. (8.1-5) and write

$$\frac{\partial^2 \psi}{\partial t^2} - v^2 \frac{\partial^2 \psi}{\partial z^2} \approx \beta_2 \frac{\partial^2 \psi^2}{\partial t^2}. \qquad (8.1\text{-}6)$$

Note that both Eqs. (8.1-5) and (8.1-6) reduce to the one-dimensional homogeneous wave equation if $\beta_2 = 0$. For the more general case $\beta_2 \neq 0$, $\beta_3 \neq 0$, a similar manipulation gives

$$\frac{\partial^2 \psi}{\partial t^2} - v^2 \frac{\partial^2 \psi}{\partial z^2} \approx \beta_2 \frac{\partial^2 \psi^2}{\partial t^2} + \frac{2}{3}\beta_3 \frac{\partial^2 \psi^3}{\partial t^2}. \qquad (8.1\text{-}7)$$

We can extend Eq. (8.1-7) to higher dimensions by replacing $\partial^2/\partial z^2$ by the Laplacian ∇^2,

$$\frac{\partial^2 \psi}{\partial t^2} - v^2 \nabla^2 \psi \approx \beta_2 \frac{\partial^2 \psi^2}{\partial t^2} + \frac{2\beta_3}{3}\frac{\partial^2 \psi^3}{\partial t^2}. \qquad (8.1\text{-}8)$$

Here, too, notice that β_2 and β_3 appear with the terms involving ψ^2 and ψ^3, respectively, justifying the names quadratic and cubic nonlinearity coefficients given previously.

As will become clear later in this section, Eq. (8.1-8) with $\beta_2 \neq 0$ and $\beta_3 \neq 0$ becomes the wave equation for a component of the **E** field, where we identify the RHS as the source term due to the nonlinear polarization of the medium. The quadratic and cubic components of the nonlinear induced polarization are proportional to $\beta_2 \partial^2 \psi^2/\partial t^2$ and $(2\beta_3/3)\partial^2 \psi^3/\partial t^2$, respectively.

At this point, it is instructive to strike a connection between the nonlinearity parameters β_2 and β_3 introduced previously and the commonly occurring parameters in the nonlinear optics litera-ture. Conventionally, the description of nonlinearities is given in terms of the nonlinear induced polarization in the medium or of an amplitude-dependent refractive index. In the former case, the in-duced polarization **P**, having components P_i, is expressed in terms of a Taylor series expansion of the electric field component E_j according to [Yariv and Yeh (1984)]

$$P_i = \epsilon_0 \chi_{ij} E_j + \epsilon_0 \chi_{ijk} E_j E_k + \epsilon_0 \chi_{ijkl} E_j E_k E_l + \cdots \qquad (8.1\text{-}9a)$$

$$\triangleq \epsilon_0 \chi_{ij} E_j + 2d_{ijk} E_j E_k + \epsilon_0 \chi_{ijkl} E_j E_k E_l + \cdots \qquad (8.1\text{-}9b)$$

$$\triangleq \epsilon_0 \chi_{ij} E_j + \epsilon_0 P_i^{\text{NL}}, \qquad (8.1\text{-}9c)$$

where we have used the Einstein convention for summation, intro-duced in Chapter 1. While the χ_{ij}s denote the *linear susceptibility tensor*, the χ_{ijk}s (or d_{ijk}s) and χ_{ijkl}s represent the *nonlinear suscep-tibility tensors*. In Eq. (8.1-9c), P_i^{NL} refers to the nonlinear part of the polarization P_i. We remark, however, that this tensor representation is not usually necessary to explain the physics of the nonlinear effects mentioned earlier. However, readers should be familiar with it for cross-referencing with existing literature.

We now derive the wave equation for the vector electric field $\mathbf{E}$ in terms of the nonlinear polarization $\mathbf{P}^{NL}$, whose components are written as P_i^{NL}. From Maxwell's equations [Eqs. (3.1-1)–(3.1-4)] and using

$$\mathbf{D} = \epsilon_0 \mathbf{E} + \mathbf{P}, \qquad (8.1\text{-}10)$$

we have

$$\mathbf{\nabla} \times \mathbf{\nabla} \times \mathbf{E} = \mathbf{\nabla}(\mathbf{\nabla} \cdot \mathbf{E}) - \nabla^2 \mathbf{E} = -\mu_0 \epsilon_0 \frac{\partial^2 \mathbf{E}}{\partial t^2} - \mu_0 \frac{\partial^2 \mathbf{P}}{\partial t^2}, \quad (8.1\text{-}11)$$

similar to the way we derived the wave equation for linear homogeneous isotropic media in Chapter 3. There, for a source-free region, we argued that $\mathbf{\nabla} \cdot \mathbf{E} = 0$ because $\mathbf{\nabla} \cdot \mathbf{D} = 0$. However, note from our discussion of linear wave propagation in crystals (which are anisotropic) in Chapter 6 that, for time-harmonic fields, $\mathbf{D}$ is proportional to the component $\mathbf{E}_\perp$ of the total electric field which is transverse to the direction of propagation. Only if we assume that the total electric field is entirely transverse, that is, $\mathbf{E} = \mathbf{E}_\perp$, does $\mathbf{D}$ become proportional to $\mathbf{E}$. For this case,

$$\frac{\partial^2 \mathbf{E}}{\partial t^2} - \frac{1}{\mu_0 \epsilon_0} \nabla^2 \mathbf{E} = -\frac{1}{\epsilon_0} \frac{\partial^2 \mathbf{P}}{\partial t^2}. \qquad (8.1\text{-}12)$$

Now, on the basis of Eq. (8.1-9c),

$$\mathbf{P} = \mathbf{P}^L + \mathbf{P}^{NL}. \qquad (8.1\text{-}13)$$

If the medium is isotropic in the linear regime, that is, $\chi_{ij} = \chi$, where χ denotes the linear susceptibility,

$$\mathbf{P} = \epsilon_0 \chi \mathbf{E} + \mathbf{P}^{NL}, \qquad (8.1\text{-}14)$$

and Eq. (8.1-12) reduces to

$$\frac{\partial^2 \mathbf{E}}{\partial t^2} - v^2 \nabla^2 \mathbf{E} = -\frac{1}{\epsilon} \frac{\partial^2 \mathbf{P}^{NL}}{\partial t^2}, \qquad (8.1\text{-}15)$$

where

$$\epsilon = \epsilon_0(1 + \chi), \qquad v^2 = \frac{1}{\mu_0 \epsilon}. \qquad (8.1\text{-}16)$$

It is true that through Eq. (8.1-16) we can obtain a set of coupled PDEs, in general, that describes the behavior of a certain component of the electric field. In many cases of practical interest, however, it may turn out that only one component of the electric field $\mathbf{E}$, namely, E_1, is present in the nonlinear medium. In that case, only χ_{111} (or d_{111}) and χ_{1111} are the pertinent quadratic and cubic nonlinearity constants, and Eq. (8.1-15) reduces to an equation of the same form as Eq. (8.1-8),

$$\frac{\partial^2 E_1}{\partial t^2} - v^2 \nabla^2 E_1 = -\frac{2}{\epsilon} \frac{\partial^2}{\partial t^2} \left[d_{111} E_1^2 + \epsilon_0 \chi_{1111} E_1^3 \right]. \quad (8.1\text{-}17)$$

If more than one component of the electric field, namely, E_1 and E_2, are present, the nonlinearly coupled PDEs describing E_1 and E_2, using Eq. (8.1-15), become

$$\frac{\partial^2 E_1}{\partial t^2} - v^2 \nabla^2 E_1 = -\frac{2}{\epsilon} \frac{\partial^2}{\partial t^2} \left[d_{111} E_1^2 + (d_{112} + d_{121}) E_1 E_2 + d_{122} E_2^2 \right],$$

$$\frac{\partial^2 E_2}{\partial t^2} - v^2 \nabla^2 E_2 = -\frac{2}{\epsilon} \frac{\partial^2}{\partial t^2} \left[d_{211} E_1^2 + (d_{212} + d_{221}) E_1 E_2 + d_{222} E_2^2 \right],$$

$$(8.1\text{-}18)$$

where we have considered only the quadratic nonlinearity, for simplicity. It might turn out that a certain crystal has many of its d_{ijk}s equal to zero or negligibly small in comparison to its nonzero components (viz., d_{111} and d_{222}) to facilitate decoupling between the preceding two equations. In that case, too, each equation

Table 8.1 Quadratic Nonlinearity Coefficients d_{ijk} for Some Crystals

Crystal	λ_ν (μm)	d_{ijk} ($\frac{1}{9} \times 10^{-22}$ m^3 / V-s^2)
GaAs	10.6	$d_{123} : 107$
InSb	28	$d_{123} : 462$
CdTe	28	$d_{123} : 48$
CdS	10.6	$d_{333} : 35; \ d_{311} : -21$
	1.06	$d_{333} : 80$
LiNbO$_3$	1.06	$d_{333} : -27; \ d_{222} : 4$
Quartz	1.064	$d_{111} : 0.4$
KDP	0.6328	$d_{213} : 0.57$
	1.06	$d_{312} : 0.50$

reduces to a form similar to Eq. (8.1-8). The most general case is quite complicated to tackle and is outside the scope of this book.

In view of the preceding discussion, we will use Eq. (8.1-3) or (8.1-8) to explain the essential physics behind the nonlinear effects mentioned in the beginning of this section.

To obtain typical values for β_2, note that, for instance, a comparison of Eqs. (8.1-8) (with $\beta_3 = 0$) and (8.1-17) (with $\chi_{1111} = 0$) shows that $\beta_2 = -(2/\epsilon)d_{111}$. More generally, we can say that

$$\beta_2 \sim -\frac{2}{\epsilon} d_{ijk}, \qquad (8.1\text{-}19)$$

where $\sim$ means "of the order of." Table 8.1 lists some typical values for d_{ijk} along with the corresponding wavelengths at which they are measured [Yariv and Yeh (1984)].

Now, to prescribe typical values for β_3, note that a comparison of Eqs. (8.1-8) (with $\beta_2 = 0$) and (8.1-17) (with $d_{ijk} = 0$) shows that

$$\beta_3 = -3\frac{\epsilon_0}{\epsilon}\chi_{1111}. \qquad (8.1\text{-}20)$$

Values of χ_{1111} for different materials are listed in Table 8.2 [Yariv (1985)]. Often, the cubic nonlinearity coefficient is specified in terms of other parameters, such as the *nonlinear refractive index coefficient*. One of the definitions for this is through the relation

$$n(x, y, z) = n + \tfrac{1}{2}n_2\psi_e\psi_e^*, \qquad (8.1\text{-}21)$$

Table 8.2 Cubic Nonlinearity Coefficients χ_{1111} for Some Materials

Material	λ_ν (μm)	n	χ_{1111} $[(4\pi/9) \times 10^{-21}$ m^2/V$^2]$
GaAs	10.6	3.3	120
InSb	5.3	4	-6×10^{10}
	10.6	4	2×10^6
CdTe	1.06	3	2.5×10^5
CdS	0.694	2.42	130
Ge	10.6	4	10^3
Si	10.6	3.4	60
H$_2$O	0.694	1.33	0.7
Acetone	0.694	1.35	1.8
Benzene	1.064	1.5	2.4

where $\psi_e = \psi_e(x, y, z)$ represents the envelope of $\psi(x, y, z, t) = \text{Re}\{\psi_e(x, y, z)\exp[j(\omega_0 t - k_0 z)]\}$; $n(x, y, z)$ is the overall refractive index, n denotes the linear refractive index, and n_2 is the nonlinear refractive index coefficient. We can relate n_2 with β_3 by starting from Eq. (8.1-2) with $\beta_2 = 0$ and assuming weak nonlinearity to get

$$n(x, y, z) = n - n\beta_3\langle\psi\rangle^2. \qquad (8.1\text{-}22)$$

The time average $\langle\ \rangle$ has been inserted to represent the fact that $n(x, y, z)$ is only dependent on the *slowly varying* part of ψ. Next, realizing that $\langle\psi\rangle^2 = \frac{1}{2}\psi_e\psi_e^*$, we get, upon comparing Eq. (8.1-21) and (8.1-22), the relation

$$\beta_3 = -\frac{n_2}{n}. \qquad (8.1\text{-}23)$$

The units of n_2 and β_3 are m^2/V^2 in MKS since ψ_e has units of V/m. We specifically point this out for the benefit of readers who will no doubt be frustrated at different unit systems that are used in literature, as well as different definitions of the nonlinear polarization and the nonlinear refractive index coefficient. For instance, an equivalent representation of the nonlinear refractive index that appears in the literature is [Haus (1984)]

$$n(x, y, z) = n + \gamma I, \qquad (8.1\text{-}24a)$$

with

$$I = I(x, y, z) = \epsilon v \frac{|\psi_e|^2}{2}, \qquad (8.1\text{-}24b)$$

where I is the optical intensity or irradiance, introduced in Chapter 3 [see Eq. (3.2-33)]; γ is in units of m^2/W and is related to n_2 as

$$\gamma = \frac{n_2}{\epsilon v}. \qquad (8.1\text{-}25)$$

What is the physical origin of nonlinear polarization? To answer this question, recall that in Chapter 3 (Section 3.2.4), we briefly discussed electronic polarization, which contributes to dispersion. This happens when the frequency of the applied electric field is close to the resonant frequency of the atom. The resulting polarization is linearly proportional to the electric field and can be derived by equating the total force on an electron, comprising the electrical force and the mechanical force, to zero. Now, the mechanical force in the linear approximation is proportional to the displacement between the electron and the nucleus of the atom, much like the restoring force of a spring in a classical spring–mass problem. For larger displacements, however, the mechanical force is a *nonlinear* function of the displacement and can be approximately expressed in terms of a power series. If we equate the forces on the electron once again, we realize that the displacement and hence the polarization, which is proportional to it, is a nonlinear function of the applied electric field and can also be expressed in terms of a power series. We refer interested readers to Yariv (1985) for a more detailed treatment.

8.2 Harmonic and Subharmonic Generation

An important consequence of nonlinearity is frequency multiplication. To see this, consider, first, one-dimensional continuous-wave (CW) propagation in a quadratically nonlinear ($\beta_2 \neq 0$, $\beta_3 = 0$) nondispersive medium. If a CW wave of angular frequency ω_0 and propagation constant k_0 is incident on such a medium, it can generate harmonics that travel along with the same velocity. The *energy* and *momentum conservation laws* as well as the *dispersion relationship* (variation of ω with k) have to be satisfied whenever wave interactions occur [Phillips (1974)]. Limiting ourselves to only two frequencies, namely, the *fundamental* (ω_0, k_0) and the *second harmonic* $(2\omega_0, 2k_0)$, the following interactions occur:

$$\omega_0 + \omega_0 \rightarrow 2\omega_0, \qquad 2\omega_0 - \omega_0 \rightarrow \omega_0,$$

$$k_0 + k_0 \rightarrow 2k_0, \qquad 2k_0 - k_0 \rightarrow k_0,$$

(8.2-1)

with $\omega_0/k_0 = v$, the phase velocity of the two propagating frequencies. Relations as in (8.2-1) can be understood by examining the square of the sum of two sinusoidal functions, namely, $A_1 \cos \omega_0 t + A_2 \cos 2\omega_0 t$. The result contains terms proportional to $A_1^2 \cos 2\omega_0 t$ and $A_1 A_2 \cos \omega_0 t$. Loosely speaking, the presence of A_1^2 in the first term tells us that a frequency $2\omega_0$ is created from the self-interaction of the wave at frequency ω_0, while the occurrence of $A_1 A_2$ in the second term implies that energy is transferred back to the frequency ω_0 due to interaction between waves at frequencies $2\omega_0$ and ω_0. Because the preceding results were derived by *squaring* the time function, the quadratic nonlinearity (β_2) is responsible for this sort of energy exchange. This is the basis of the second-harmonic generation usually observed in nonlinear electrooptic crystals [Bloembergen (1965)].

The interactions represented by Eqs. (8.2-1) also describe *subharmonic* generation. In this case, energy at a frequency $2\omega_0$ (now considered as the fundamental) is transferred to its subharmonic (at frequency ω_0). However, as the second interaction depicted in Eqs. (8.2-1) suggests, a finite amount of energy must be present initially at ω_0 to initiate the subharmonic generation. In practice, this is always the case, because there is noise at all frequencies, omnipresent in any system. Thus, the physics of subharmonic generation is fundamentally different from the second-harmonic generation process.

The next question to ask is: Could a cubic nonlinearity also play a role in second-harmonic generation? At first thought, this seems doubtful, because a cubic nonlinearity can only facilitate interactions of the type $\omega_0 + \omega_0 + \omega_0 \rightarrow 3\omega_0$ leading to third-harmonic generation. A closer examination reveals, however, that the answer is yes, *provided the quadratic nonlinearity is also present initially to generate the second harmonic*. In this case, we can represent the frequency mixings that occur as

$$2\omega_0 + \omega_0 - \omega_0 \to 2\omega_0, \qquad 2\omega_0 - 2\omega_0 + 2\omega_0 \to 2\omega_0,$$

$$\omega_0 + 2\omega_0 - 2\omega_0 \to \omega_0, \qquad \omega_0 - \omega_0 + \omega_0 \to \omega_0,$$

$$(8.2\text{-}2)$$

with similar relations for the k's. We can visualize the parametric process described above as additional contributions to the fundamental and second-harmonic amplitudes in the presence of the cubic nonlinearity. Once again, relations as in (8.2-2) can be understood by *cubing* the function $A_1 \cos \omega_0 t + A_2 \cos \omega_0 t$, and the corresponding energy exchange is effected through the cubic nonlinearity (β_3). A detailed discussion of this is outside the scope of this book.

If, now, the medium under consideration is *dispersive* (see Section 3.2.4), the propagation constants for the frequencies ω_0 and $2\omega_0$ are no longer related by a simple factor [viz., $k_2 \triangleq k(2\omega_0) = 2k_1 \triangleq 2k(\omega_0)$]. In nonlinear optics, this situation is termed the *phase-mismatched case*. For one-dimensional propagation, this gives rise to a *spatial beating* for the fundamental and harmonic (subharmonic) amplitudes, which in the general theory of nonlinear waves is called the *Fermi–Pasta–Ulam* (FPU) *recurrence* phenomenon. It is customary to assess the amount of dispersion by computing the *recurrence period* which is defined as the length two propagating frequencies ω_0 and $2\omega_0$, initially in phase, travel to get back in phase again. In the linear case, this is given by $2\pi / |(k_2 - 2k_1)|$. However, as we shall see in Section 8.2.2, the recurrence period for nonlinear dispersive propagation is slightly modified by the presence of nonlinearities. In nonlinear optics literature, it is more conventional to refer to this distance as the coherence length instead of the recurrence period. The *coherence length* l_c is a measure of the maximum length of the material that is useful in producing the second-harmonic power, and is equal to one-half the recurrence period. We can reexpress l_c as $\lambda_v / 4|(n^{2\omega_0} - n^{\omega_0})|$, where λ_v is the wavelength in vacuum and n^{ω_0} and $n^{2\omega_0}$ are the refractive indices at frequencies ω_0 and $2\omega_0$, respectively.

In many physical situations, however, wave propagation is essentially higher-dimensional unless special care is taken to ensure one-dimensional propagation. For the higher-dimensional disper-

Nonlinear Optics

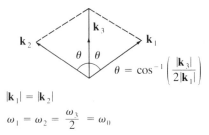

$$|\mathbf{k}_1| = |\mathbf{k}_2|$$

$$\omega_1 = \omega_2 = \frac{\omega_3}{2} = \omega_0$$

Figure 8.2 Resonant-triad wave-vector diagram for three interacting waves, two of which have the same temporal frequency.

sive case, assuming quadratic nonlinearity, the two frequencies are no longer constrained to propagate in the same direction and hence interact according to the *resonant-triad* relations [Phillips (1974)]

$$\omega_1 + \omega_2 \rightarrow \omega_3, \qquad \mathbf{k}_1 + \mathbf{k}_2 \rightarrow \mathbf{k}_3 \qquad \text{(8.2-3a)}$$

with

$$\omega_i = \omega(|\mathbf{k}_i|). \qquad \text{(8.2-3b)}$$

The resonance triad vector diagram is shown in Figure 8.2 for the case $\omega_1 = \omega_2 = \omega_0$ and $\omega_3 = 2\omega_0$. Because the waves are free to travel in higher dimensions (predictable from the vector diagram and the dispersion relationship), there is no longer any spatial beating due to dispersion, and the evolution of the second harmonic (or the subharmonic) is similar to the nondispersive case, albeit in their respective directions. The rigorous mathematical formulation of this case is beyond the scope of this discussion.

8.2.1 Mathematical Formulation of One-Dimensional Propagation

In what follows, we analyze second-harmonic (and subharmonic) generation by using the simple model of Eqs. (8.1-3) and by heuristically incorporating the effect of dispersion (phase mismatch). We assure readers that the results are identical to those derived in existing nonlinear optics literature that use Eq. (8.1-15) as the starting point.

We first represent $\psi(z, t)$ as

$$\psi(z, t) = \tfrac{1}{2} \sum_{n=-2}^{2} \Psi_n(z) \exp[jn(\omega_0 t - k_0 z)], \qquad (8.2\text{-}4)$$

with $\Psi_0 = 0$ (we assume no dc component), and with $\Psi_{-n} = \Psi_n^*$ to ensure that the wavefunction is real. The reason we terminate the series as in Eq. (8.2-4) is that we are only interested in the evolution of the spectral amplitudes Ψ_n of ω_0 and $2\omega_0$ when light at frequency ω_0 (or $2\omega_0$) is incident on a quadratically nonlinear medium. We next substitute Eq. (8.2-4) into Eq. (8.1-3a) with $\beta_3 = 0$. This gives

$$\sum_{n=-2}^{2} \left[\frac{d\Psi_n(z)}{dz} \right] \exp[jn(\omega_0 t - k_0 z)]$$

$$+ \frac{\beta_2}{2} \left\{ \frac{\partial}{\partial z} \left(\sum_{n=-2}^{2} \Psi_n(z) \exp[jn(\omega_0 t - k_0 z)] \right) \right\}$$

$$\times \left(\sum_{m=-2}^{2} \Psi_m(z) \exp[jm(\omega_0 t - k_0 z)] \right) = 0, \qquad (8.2\text{-}5)$$

where we have assumed $\omega_0/k_0 = v$. We rewrite the product of two series in (8.2-5) using *Cauchy's product rule*,

$$\sum_n A_n \sum_m B_m = \sum_n \sum_m A_m B_{n-m}. \qquad (8.2\text{-}6)$$

The last term on the left-hand side of (8.2-5) thus becomes

$$-j\frac{\beta_2}{2} \sum_{n=-2}^{2} \sum_{m=-2}^{2} mk_0 \Psi_m \Psi_{n-m} \exp[jn(\omega_0 t - k_0 z)], \qquad (8.2\text{-}7)$$

where we have employed the *slowly varying envelope approximation* (SVEA) of Ψ_n,

$$\left| \frac{d\Psi_n}{dz} \right| \ll nk_0 |\Psi_n|. \qquad (8.2\text{-}8)$$

Using Eq. (8.2-7) in Eq. (8.2-5) and equating the coefficients of $\exp[jn(\omega_0 t - k_0 z)]$, we obtain the evolution equation for the spectral components Ψ_n,

$$\frac{d\Psi_n}{dz} = j\frac{k_0 \beta_2}{2} \sum_{m=-2}^{2} m\Psi_m \Psi_{n-m}, \qquad (8.2\text{-}9)$$

where the term on the right-hand side denotes the contribution from the quadratic nonlinearity. In practice, higher harmonics are generated, and their evolution can be analyzed through replacing the limits of the summations in Eq. (8.2-9) by $\pm\infty$. For most physical systems, however, contributions from frequencies higher than ω_0 and $2\omega_0$ are not dominant. Furthermore, higher harmonics decay faster than lower harmonics.

Now, a physical system may also be dispersive. To account for the effect of dispersion, we must slightly modify Eq. (8.2-9). In the presence of dispersion, the propagation constant for the nth harmonic $n\omega_0$ is no longer nk_0, but can be written as $k_n \neq nk_0$. Thus, during linear propagation in a dispersive medium, the spatial behavior of Ψ_n will be

$$\Psi_n(z) = \Psi_n(0)\exp[-j(k_n - nk_0)z], \qquad (8.2\text{-}10)$$

so that in Eq. (8.2-4),

$$\psi(z, t) = \tfrac{1}{2} \sum_{n=-2}^{2} \Psi_n(0)\exp[j(n\omega_0 t - k_n z)], \qquad (8.2\text{-}11)$$

as expected. From Eq. (8.2-10),

$$\frac{d\Psi_n}{dz} = -j(k_n - nk_0)\Psi_n. \qquad (8.2\text{-}12)$$

To incorporate the effect of dispersion in our nonlinear system, Eq. (8.2-9), we can *heuristically* add a term, as on the right-hand side of Eq. (8.2-12), to the right-hand side of Eq. (8.2-9) *if the nonlinear and dispersive effects are small*:

$$\frac{d\Psi_n}{dz} = j\frac{k_0 \beta_2}{2} \sum_{m=-2}^{2} m\Psi_m \Psi_{n-m} - j(k_n - nk_0)\Psi_n. \quad (8.2\text{-}13)$$

Equation (8.2-13) will serve as our model equation describing the one-dimensional evolution of the fundamental and the second harmonic in a medium with quadratic nonlinearity and dispersion. We remark that this equation is essentially identical to that in the existing literature [see, for instance, Bloembergen (1965)]. Here, we will not get into an involved discussion to bring out the similarities. In what follows, we analyze second-harmonic generation in a medium with quadratic nonlinearity without and with dispersion. Comments on subharmonic generation are also made.

8.2.2 Second-Harmonic and Subharmonic Generation

Restricting ourselves to two frequencies ω_0 and $2\omega_0$ (i.e., taking $n = 1, 2$), we get the following two coupled equations from Eq. (8.2-13):

$$\frac{d\Psi_1}{dz} = j\frac{\beta_2 k_0}{2}\Psi_{-1}\Psi_2 - j(k_1 - k_0)\Psi_1, \qquad (8.2\text{-}14a)$$

$$\frac{d\Psi_2}{dz} = j\frac{\beta_2 k_0}{2}\Psi_1^2 - j(k_2 - 2k_0)\Psi_2, \qquad (8.2\text{-}14b)$$

where Ψ_1 and Ψ_2 denote the complex spectral amplitudes at ω_0 and $2\omega_0$ and where $\Psi_{-1} = \Psi_1^*$. Note that at this point, we have retained the symbols k_1 and k_2 to symbolize the propagation constants at the frequencies ω_0 and $2\omega_0$, respectively. We now resolve the complex spectral amplitudes into their respective (real) amplitudes and phases by writing

$$\Psi_n(z) = a_n(z)\exp[-j\phi_n(z)] \qquad (8.2\text{-}15)$$

in Eqs. (8.2-14a) and (8.2-14b). This yields the following set of four equations:

$$\frac{da_1}{dz} = \frac{\beta_2 k_0}{2}a_1 a_2 \sin(\phi_2 - 2\phi_1), \qquad (8.2\text{-}16a)$$

$$\frac{da_2}{dz} = -\frac{\beta_2 k_0}{2}a_1^2 \sin(\phi_2 - 2\phi_1), \qquad (8.2\text{-}16b)$$

$$\frac{d\phi_1}{dz} = -\frac{\beta_2 k_0}{2}a_2 \cos(\phi_2 - 2\phi_1) + (k_1 - k_0), \qquad (8.2\text{-}17a)$$

$$\frac{d\phi_2}{dz} = -\frac{\beta_2 k_0}{2}\frac{a_1^2}{a_2} \cos(\phi_2 - 2\phi_1) + (k_2 - 2k_0). \qquad (8.2\text{-}17b)$$

We combine Eqs. (8.2-17a) and (8.2-17b) by defining

$$\theta \triangleq \phi_2 - 2\phi_1 \qquad (8.2\text{-}18)$$

to give

$$\frac{d\theta}{dz} = \Delta k + \frac{\beta_2 k_0}{2}\left(2a_2 - \frac{a_1^2}{a_2}\right)\cos\theta, \qquad (8.2\text{-}19)$$

where

$$\Delta k \triangleq k_2 - 2k_1 \qquad (8.2\text{-}20)$$

is related to the *phase velocity mismatch* due to dispersion. We remark that, from Eqs. (8.2-16a) and (8.2-16b), it readily follows that

$$a_1^2(z) + a_2^2(z) = \text{constant} = a_1^2(0) + a_2^2(0) = \tilde{E}, \quad (8.2\text{-}21)$$

depicting conservation of energy. Finally, normalization using the definitions

$$\xi = \frac{\beta_2 k_0}{2}\tilde{E}^{1/2}z, \qquad \Delta s = \frac{2\,\Delta k}{\beta_2 k_0 \tilde{E}^{1/2}},$$

$$u = \frac{a_1}{\tilde{E}^{1/2}}, \qquad w = \frac{a_2}{\tilde{E}^{1/2}}, \qquad (8.2\text{-}22)$$

reduces the system of equations, Eqs. (8.2-16a), (8.2-16b), and (8.2-19), to

$$\frac{du}{d\xi} = uw\sin\theta, \qquad (8.2\text{-}23a)$$

$$\frac{dw}{d\xi} = -u^2\sin\theta, \qquad (8.2\text{-}23b)$$

$$\frac{d\theta}{d\xi} = \Delta s + \left(2w - \frac{u^2}{w}\right)\cos\theta, \qquad (8.2\text{-}24)$$

where we have used the definition of θ as in Eq. (8.2-18). Observe, also, that using Eqs. (8.2-23a) and (8.2-23b),

$$\frac{d\left[\ln(u^2 w)\right]}{d\xi} = \frac{2}{u}\frac{du}{d\xi} + \frac{1}{w}\frac{dw}{d\xi}$$

$$= \left(2w - \frac{u^2}{w}\right)\sin\theta. \qquad (8.2\text{-}25)$$

Using Eq. (8.2-25), we can recast Eq. (8.2-24) in the form

$$\frac{d\theta}{d\xi} = \Delta s + \cot\theta\frac{d\left[\ln(u^2 w)\right]}{d\xi}. \qquad (8.2\text{-}26)$$

Equations (8.2-23a), (8.2-23b), and (8.2-26) describe the evolution of the real amplitudes of the waves at frequencies ω_0 and $2\omega_0$, and the relative phase difference during propagation in a quadratically nonlinear dispersive medium [Bloembergen (1965) and Banerjee and Korpel (1981)]. In what follows, we analyze second-harmonic generation and discuss subharmonic generation using Eqs. (8.2-23a), (8.2-23b), and (8.2-26) as our starting point.

Case A: Perfect phase matching $\Delta s = 0$.
From Eq. (8.2-26), one integration yields, after straightforward algebra,

$$u^2 \cos\theta = \text{constant} \triangleq \Upsilon = u^2(0)w(0)\cos\theta(0), \quad (8.2\text{-}27)$$

where $u(0)$, $w(0)$ and $\theta(0)$ are the values of u, w and θ at $\xi = 0$. Now, from Eqs. (8.2-21) and (8.2-22), it follows that

$$u^2 + w^2 = 1. \qquad (8.2\text{-}28)$$

Using Eqs. (8.2-23b), (8.2-27), and (8.2-28), we can write

$$\frac{d(w^2)}{d\xi} = \mp 2w(1 - w^2)\left[1 - \frac{\Upsilon^2}{w^2}(1 - w^2)^2\right]^{1/2}$$

or

$$\xi = \pm\frac{1}{2}\int_{w^2(0)}^{w^2(\xi)}\left[w^2(1 - w^2)^2 - \Upsilon^2\right]^{-1/2}d(w^2). \qquad (8.2\text{-}29)$$

Nonlinear Optics

The general solution to $w(\xi)$ [and hence to $u(\xi)$ and $\theta(\xi)$], as given by Eq. (8.2-29), is rather complicated if $w(0) \neq 0$. Suffice it here to state that the solution depends on the roots of the equation

$$w^2(1 - w^2)^2 - \Upsilon^2 = 0, \qquad (8.2\text{-}30)$$

which, for now, we write as w_a^2, w_b^2, and w_c^2, with

$$w_c^2 \geq w_b^2 \geq w_a^2. \qquad (8.2\text{-}31)$$

The solution to Eq. (8.2-29) is expressible in terms of Jacobian elliptic functions like "cn" and "sn" that look similar to the cosine and sine functions, and are periodic in nature. The period Π_ξ of $v(\xi)$ is given as [Abramowitz and Stegun (1965)]

$$\Pi_\xi = \int_{w_a^2}^{w_b^2} \left[w^2(1 - w^2)^2 - \Upsilon^2 \right]^{-1/2} d(w^2). \qquad (8.2\text{-}32)$$

A case of physical interest is where $w(0) = 0$ [$u(0) \neq 0$], that is, when there is no initial second harmonic in the system. Then $\Upsilon = 0$ [from Eq. (8.2-27)] and $w_a^2 = 0$ and $w_b^2 = w_c^2 = 1$ [from Eqs. (8.2-30) and (8.2-31)]. The period Π_ξ [using Eq. (8.2-32)] becomes infinity. We can then easily integrate Eq. (8.2-29) to get

$$w(\xi) = \tanh \xi. \qquad (8.2\text{-}33a)$$

Then from Eq. (8.2-28),

$$u(\xi) = \operatorname{sech} \xi. \qquad (8.2\text{-}33b)$$

Note that using Eqs. (8.2-23a) or (8.2-23b), (8.2-33a), and (8.2-33b), $\theta(0) = -\pi/2$. From Eq. (8.2-23b), it follows that $\theta(\xi) = \theta(0) = -\pi/2$, i.e., the relative phase difference remains locked at the initial value.

A plot of u and w is shown in Figure 8.3(a). Observe that the fundamental amplitude asymptotically goes to 0, whereas the normalized second-harmonic amplitude asymptotically tends to 1. In a material of finite length L, the amount of second harmonic generated is usually described in terms of the *conversion efficiency*

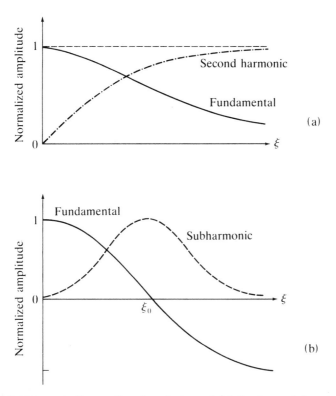

Figure 8.3 Theoretically predicted variation of (a) fundamental and second-harmonic amplitudes and (b) fundamental and subharmonic amplitudes due to quadratic nonlinearity and in the case of perfect phase matching.

η_{SHG}. This is defined as the ratio of the power in the second harmonic at $z = L$, and the initial fundamental power. Using Eqs. (8.2-22) and (8.8-33a), this becomes

$$\eta_{SHG} = \tanh^2 \frac{\beta_2 k_0 a_1(0) L}{2}, \qquad (8.2\text{-}34)$$

where $a_1(0)$ is the initial fundamental amplitude.

We remark, in passing, that Eqs. (8.2-23a), (8.2-23b), and (8.2-24) can also be used to explain subharmonic generation. For this, we have to take $w(0) \neq 0$ and assume that a small amount of the subharmonic (u) exists in the system due to ambient noise. If we assume this initial amount of the subharmonic [$= \bar{\epsilon}w(0)$, where $\bar{\epsilon}$ is a small quantity], then for $\theta(0) = \pi/2$, the solutions to Eqs. (8.2-23a) and (8.2-23b) are

$$u(\xi) = \text{sech}(\xi - \xi_0), \qquad (8.2\text{-}35a)$$

$$w(\xi) = -\tanh(\xi - \xi_0). \qquad (8.2\text{-}35b)$$

In Eqs. (8.2-35a) and (8.2-35b),

$$\xi_0 = \tanh^{-1} w(0) \approx \frac{1}{w(0)} \ln\left(\frac{1}{\epsilon}\right) \qquad (8.2\text{-}36)$$

with

$$w^2(0)(1 + \bar{\epsilon}^2) = 1, \qquad (8.2\text{-}37)$$

to ensure conservation of energy. A plot of u and w for this case is shown in Figure 8.3(b). The details of the derivation of Eqs. (8.2-35a) and (8.2-35b) are left as an exercise for the readers.

Case B: Imperfect phase matching $\Delta s \neq 0$.

Multiplying Eq. (8.2-24) with $u^2 w \sin \theta$ and using Eqs. (8.2-23a) and (8.2-23b), we get [Bloembergen (1965), and Banerjee and Korpel (1981)]

$$\frac{d[u^2 w \cos \theta]}{d\xi} + \frac{\Delta s}{2} \frac{d(w^2)}{d\xi} = 0. \qquad (8.2\text{-}38)$$

Integration gives

$$u^2 w \cos \theta + \frac{\Delta s}{2} w^2 = \text{constant} \triangleq \Upsilon_{\Delta s} = \Upsilon + \frac{\Delta s}{2} w^2(0). \quad (8.2\text{-}39)$$

Equation (8.2-29) is now generalized as

$$\xi = \pm \frac{1}{2} \int_{w^2(0)}^{w^2(\xi)} \left[w^2(1 - w^2)^2 - \left\{ \Upsilon - \frac{\Delta s}{2} (w^2 - w^2(0)) \right\}^2 \right]^{-1/2} d(w^2).$$

(8.2-40)

Everything previously said about Eq. (8.2-29) and its solutions for $w(0) \neq 0$ carries over to the solution of Eq. (8.2-40), and the period in this case is given by

$$\Pi_\xi = \int_{w_a^2}^{w_b^2} \left[w^2(1 - w^2)^2 - \left\{ \Upsilon - \frac{\Delta s}{2} (w^2 - w^2(0)) \right\}^2 \right]^{-1/2} d(w^2),$$

(8.2-41)

where w_c^2, w_b^2, and w_a^2 (with $w_c^2 \geq w_b^2 \geq w_a^2$) denote the roots of the quantity in square brackets in Eq. (8.2-41).

As before, a case of physical interest is where $w(0) = 0$ $[u(0) \neq 0]$. In this case, we observe periodic behavior of the solutions of the fundamental and the second harmonic even with zero initial energy in the second harmonic, unlike the nondispersive case. We now find an approximate estimate for the period of the oscillations. To do this, we recast Eq. (8.2-41) in the form

$$\Pi_\xi = \frac{2}{w_c} \int_0^{\pi/2} \left[1 - \left(\frac{w_b}{w_c} \right)^2 \sin^2 \alpha \right]^{-1/2} d\alpha \qquad (8.2\text{-}42)$$

by employing the substitution

$$\sin^2 \alpha = \frac{w^2 - w_a^2}{w_b^2 - w_a^2} \qquad (8.2\text{-}43)$$

and noting that for $w(0) = 0$, $\Upsilon = 0$ and $w_a = 0$. The other two roots are given by the solution to the algebraic equation

$$(1 - w^2)^2 - \left(\frac{\Delta s}{2} \right)^2 w^2 = 0 \qquad (8.2\text{-}44)$$

Nonlinear Optics

as

$$w_c^2 = \left[\frac{\Delta s}{4} + \left\{ 1 + \left(\frac{\Delta s}{4} \right)^2 \right\}^{1/2} \right]^2 = \frac{1}{w_b^2}. \qquad (8.2\text{-}45)$$

For the *severely phase-mismatched case* ($\Delta s \gg 1$), it follows from Eq. (8.2-45), after simple algebra, that

$$w_b^2 \approx \left(\frac{2}{\Delta s} \right)^2 \left[1 - \frac{8}{(\Delta s)^2} \right], \qquad (8.2\text{-}46\text{a})$$

$$w_c^2 \approx \left(\frac{\Delta s}{2} \right)^2 \left[1 + \frac{8}{(\Delta s)^2} \right] \gg w_b^2, \qquad (8.2\text{-}46\text{b})$$

so that Eq. (8.2-42) straightforwardly yields

$$\Pi_\xi = \frac{2\pi}{\Delta s} \left(1 - \frac{4}{(\Delta s)^2} \right). \qquad (8.2\text{-}47)$$

The period in z, or the coherence length l_c, can be readily found using the transformations in Eqs. (8.2-22). The growth of the second-harmonic amplitude for different values of phase mismatch is shown in Figure 8.4.

For efficient second-harmonic generation, it is clear, from Figure 8.4, that we must try to reduce the effects of phase mismatch

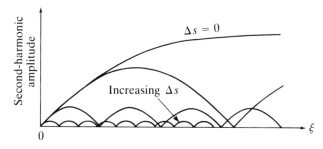

Figure 8.4 Growth of second-harmonic amplitude with propagation for the phase-mismatched case.

since the maximum achievable amount of the second harmonic generally decreases with increasing Δs. One technique that is often used takes advantage of the natural birefringence of anisotropic crystals, which was discussed in Section 6.4. Normally, in dispersive materials, n_e and n_o have similar variations with frequency, making it impossible to make $\Delta k = 0$ when both the fundamental and the second harmonic are ordinary or extraordinary. However, under certain circumstances, it may be possible to achieve *phase-matching conditions* by making the two waves to be of different types. To see this, we refer readers back to Eq. (6.4-24), which describes the variation of n_e with the angle θ between the propagation direction and the optic axis. If $n_e^{2\omega_0} < n_o^{\omega_0}$, there exists an angle θ_0 which $n_e^{2\omega_0}(\theta_0) = n_o^{\omega_0}$; thus if the fundamental (at ω_0) is introduced along θ_0 as an ordinary wave, the second harmonic (at $2\omega_0$) is generated in the same direction as an extraordinary wave. Another way of increasing the maximum efficiency of second-harmonic generation is to use the cubic nonlinearity of the material to our advantage; this, however, is outside the scope of this book.

8.3 Self-Refraction

As one of the effects of a cubic nonlinearity on wave propagation, we will first consider *self-refraction* of a beam. To get a physical picture, consider a beam with a Gaussian profile, as shown in Figure 8.5(a). If the medium is cubically nonlinear, we can describe the nonlinearity in terms of n_2, the nonlinear coefficient of the refractive index, which is related to β_3 as given in Eq. (8.1-23). If $n_2 > 0$, a region transverse to the propagation direction with greater amplitude experiences a greater refractive index than a region with lower amplitude. The result is an intensity-dependent refractive-index profile like that in graded-index optical fibers, discussed in Section 3.6. If we trace *rays*, these would appear to bend towards the axis of propagation, indicating a reduction of the beam waist size and hence an increase in the on-axis amplitude. This simple picture would suggest that the on-axis amplitude should tend to infinity; however, this does not occur because diffraction puts a limit to the minimum waist size. This is shown in Figure 8.5(b). Heuristically speaking, this makes sense because the amount of diffraction (as predictable from the angle of diffraction) depends on the ratio of the wavelength to the waist size. For an arbitrary

Nonlinear Optics

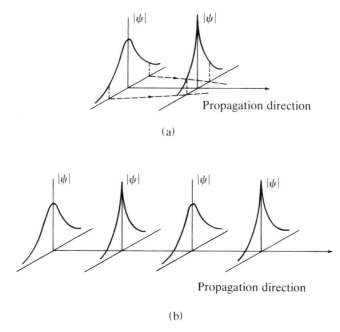

(a)

Propagation direction

(b)

Figure 8.5 Self-focussing of a Gaussian beam: (a) ray trajectory (rays tend to converge in the direction of higher refractive index); (b) periodic focussing.

initial beam profile, therefore, we would expect initial reduction of beam waist size before diffraction effects start to dominate and spread the beam again, resulting in *periodic focussing*. It turns out that although this is mostly true for a beam in a two-dimensional geometry (meaning one transverse dimension), it may not be true for the three-dimensional case. It is also possible to find the right beam profile for which the beam-narrowing effect of nonlinearity exactly balances the beam-spreading effect of diffraction [Chiao, Garmire, and Townes (1964), Haus (1966), and Akhmanov, Sukhorukov, and Khokhlov (1966)].

For the opposite kind of nonlinearity ($n_2 < 0$), it is easy to argue that the beam will spread more than it does for the linear diffraction-limited case (see Figure 8.6).

In what follows, we first derive an evolution equation for an arbitrary beam profile in a cubically nonlinear medium and then

Self-Refraction

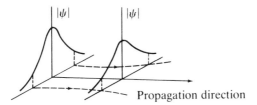

Figure 8.6 Self-focussing of a beam. The beam spreads more than in the linear diffraction-limited case.

find an analytic expression for the *diffraction-free* beam in the nonlinear medium. To this end, we express the real field $\psi(x, y, z, t)$ as

$$\psi(x, y, z, t) = \text{Re}\left[\psi_e(x, y, z)\exp(j(\omega_0 t - k_0 z))\right]. \quad (8.3\text{-}1)$$

and substitute into Eq. (8.1-8) with $\beta_2 = 0$. After straightforward algebra, we get

$$2jk_0\frac{d\psi_e}{dz} = \nabla_t^2\psi_e - \frac{\beta_3 k_0^2}{2}\psi_e^2\psi_e^*, \quad (8.3\text{-}2)$$

where we have only retained contributions around (ω_0, k_0). Furthermore, we have assumed $\omega_0/k_0 = v$ and ψ_e to be a slowly varying function of z in the sense that

$$\left|\frac{d^2\psi_e}{dz^2}\right| \ll k_0\left|\frac{d\psi_e}{dz}\right|. \quad (8.3\text{-}3)$$

In Eq. (8.3-2), ∇_t^2 denotes the transverse Laplacian $\partial^2/\partial x^2 + \partial^2/\partial y^2$. We remark that Eq. (8.3-2) with $\beta_3 = 0$ is identical to Eq. (3.3-11), which describes diffraction of a beam in a linear medium. Thus, we can regard Eq. (8.3-2) to be the nonlinear extension of Eq. (3.3-11). In Eq. (8.3-2), the first term on the right-hand side is due to diffraction, whereas the second term represents the nonlinear contribution. Equation (8.3-2) has the same form as the nonlinear Schrödinger equation [Korpel and Banerjee (1984) and Agrawal (1989)], which is used to explain soliton propagation through fibers

[Mollenauer and Stolen (1982)]. We will discuss more about soliton propagation later on in this chapter. Note that we cannot solve Eq. (8.3-2) using the Fourier transform techniques that were used to solve Eq. (3.3-11) because Eq. (8.3-2) is a nonlinear PDE.

In our quest for the expression of $|\psi_e|$ that does not depend on z, we substitute

$$\psi_e(x, y, z) = a(x, y)\exp(-j\kappa z) \qquad (8.3\text{-}4)$$

into Eq. (8.3-2) to get

$$\nabla_t^2 a = -2\kappa k_0 a + \frac{\beta_3 k_0^2}{2} a^3. \qquad (8.3\text{-}5)$$

Consider, first, the case where we have *one* transverse direction, namely, x. Equation (8.3-5) then reads

$$\frac{d^2 a}{dx^2} = -2\kappa k_0 a + \frac{\beta_3 k_0^2}{2} a^3. \qquad (8.3\text{-}6)$$

Multiplying both sides by $2\, da/dx$ and integrating, we get

$$\left(\frac{da}{dx}\right)^2 = -2\kappa k_0 a^2 + \frac{\beta_3 k_0^2}{4} a^4, \qquad (8.3\text{-}7)$$

where we have neglected the integration constant. We can recast Eq. (8.3-7) in the form

$$x = \int \frac{da}{\left[-2\kappa k_0 a^2 + (\beta_3 k_0^2/4)a^4\right]^{1/2}}. \qquad (8.3\text{-}8)$$

Equation (8.3-8) is in the form of an elliptic integral, which we first encountered while analyzing second-harmonic generation. The solution is in the form

$$a(x) = A \operatorname{sech} Kx, \qquad (8.3\text{-}9)$$

where

$$A = \left(\frac{8\kappa}{\beta_3 k_0} \right)^{1/2}, \tag{8.3-10a}$$

$$K = \frac{1}{(-2\kappa k_0)^{1/2}}. \tag{8.3-10b}$$

We note from above that $\kappa < 0$ and $\beta_3 < 0$ for a physical solution. Now $\beta_3 < 0$ implies $\varkappa_2 > 0$ [see Eqs. (8.1-23)], which is in agreement with our heuristic description for self-focussing. A plot of the beam profile is shown in Figure 8.7.

For two transverse directions, namely, x and y, we only consider the case where there is radial symmetry and express the transverse Laplacian in polar coordinates as

$$\frac{\partial^2}{\partial x^2} + \frac{\partial^2}{\partial y^2} = \frac{d^2}{dr^2} + \left(\frac{1}{r} \right) \frac{d}{dr}. \tag{8.3-11}$$

Using the definitions

$$a = \left(\frac{2\kappa}{\beta_3 k_0} \right)^{1/2} \tilde{a}, \qquad r = \left(-\frac{1}{2\kappa k_0} \right)^{1/2} \tilde{r}, \tag{8.3-12}$$

we rewrite Eq. (8.3-5) in the form

$$\frac{d^2 \tilde{a}}{d\tilde{r}^2} + \left(\frac{1}{\tilde{r}} \right) \frac{d\tilde{a}}{d\tilde{r}} - \tilde{a} + \tilde{a}^3 = 0. \tag{8.3-13}$$

This ODE has no analytic solutions; the solutions that tend to zero as $\tilde{r} \to \infty$ are obtained by numerical methods [Chiao, Garmire, and Townes (1964) and Haus (1966)]. These are shown in Figure 8.8, and are extremely sensitive to the initial condition $\tilde{a}(0)$. In a way, these solutions are reminiscent of the modes in a graded-index fiber, discussed in Section 3.6. Note that the solutions can be regarded as being *multimodal* in nature, with the mode number m depending on the initial condition $\tilde{a}(0)$.

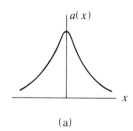

(a)

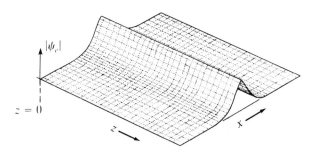

(b)

Figure 8.7 (a) Stationary nonspreading beam profile in a self-focussing medium. (b) Propagation of the nonspreading solution, numerically computed [Korpel et al. (1986)].

The solutions discussed above are *nonspreading* or *diffraction-free* beam profiles in a cubically nonlinear medium. The general solution to an arbitrary initial beam profile is difficult to obtain, since the PDEs are nonlinear. We briefly comment that an approximate solution for such an arbitrary initial beam profile can be found by starting from Eq. (8.3-2), writing down the *nonlinear eikonal equations* [similar to Eqs. (3.2-74) and (3.2-75)] by resolving ψ_e into its real amplitude and phase, and assuming the nature of solutions for these quantities [Ghatak (1978), and Banerjee, Korpel, and

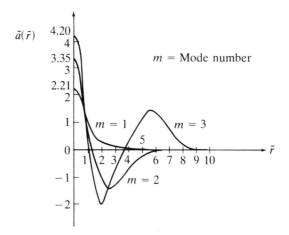

Figure 8.8 Numerical solutions to Eq. (8.3-13) showing amplitude profiles for higher-order modes.

Lonngren (1983)]. The calculations confirm periodic behavior if we consider one transverse dimension, but no periodic behavior in two transverse dimensions, for $\varkappa_2 > 0$. The detailed analysis is outside the scope of our discussion. We end this section by showing numerically computed pictures of the periodic behavior for an initial Gaussian profile (Figure 8.9) in one transverse dimension [Korpel et al. (1986)], whose amplitude and width are close to the values in (8.3-10).

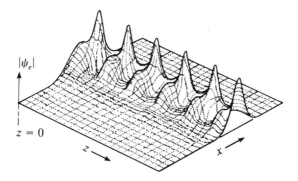

Figure 8.9 Periodic self-focussing of an initial Gaussian profile in one transverse dimension, numerically computed [Korpel et al. (1986)].

8.4 Optical Bistability: The Nonlinear Fabry – Perot Cavity

As stated in Chapter 6, optical bistability refers to the existence of two stable states of an optical system for a given set of input conditions. It is interesting physically because it represents a new kind of nonlinear system in optics. Such an effect is also of obvious practical interest, as it offers a means of realizing all-optical switching for optical computers.

In general, the requirements of these bistable characteristics require a nonlinear input–output relationship and positive feedback. As seen in Section 6.3.4, an acoustooptic device with feedback exhibits hysteresis and bistability. In this section, we consider another way of achieving optical bistability, namely, using a nonlinear Fabry–Perot cavity. An excellent reference on optical bistability is the book by Gibbs [Gibbs (1985)].

Consider a CW laser beam injected into an optical cavity that is tuned or nearly tuned to the incident light frequency. In general, the incident field is partly transmitted, partly reflected, and partly absorbed. When the cavity contains an absorbing material resonant or nearly resonant with the incident electric field, the transmitted power becomes a nonlinear function of the incident power. The behavior of the system is determined by a parameter that depends on the unsaturated absorption coefficient of the sample per unit length, on its length, and on the mirror transmittance. If one increases the value of this parameter above a certain threshold, the steady-state input–output curve becomes S-shaped, indicating bistability and hysteresis. The threshold value depends on the cavity mistuning, the atomic detuning, the inhomogeneous line width, and the type of cavity, to name a few. When the incident field is in perfect resonance with the atomic line so that dispersion does not play a role, we have *purely absorptive bistability*. On the other hand, if the atomic detuning is so large so that absorption becomes negligible, we have *purely dispersive bistability* [Lugiato (1984)].

In order to describe theoretically the phenomenon of bistability, we consider a unidirectional ring cavity as shown in Figure 8.10. For simplicity, we assume that mirrors 3 and 4 have 100% reflectivity. We denote the intensity reflection and transmission coefficients of mirrors 1 and 2 as R and T, respectively.

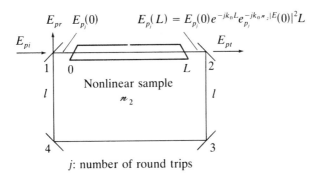

Figure 8.10 Nonlinear Fabry–Perot cavity modelled by a ring cavity. The nonlinear material is between mirrors 1 and 2.

In what follows, we will develop a simple theory of dispersive optical bistability, without taking recourse to the complicated Maxwell–Bloch equations that form the starting point of a rigorous analysis. Suffice it to state that in the dispersive case, the propagation constant in the nonlinear material is purely real but dependent on the intensity of the wave. In this case, the nonlinear material is said to behave as a *Kerr* medium. To determine the optical field inside the unidirectional ring cavity of Figure 8.10, we can write

$$E_{p_{j+1}}(0) = T^{1/2}E_{pi} + R \exp\left[-jk_0(2L + 2l)\right]$$

$$\times \exp\left[-jk_0 n_2 |E_{p_j}(0)|^2 L\right]E_{p_j}(0), \qquad (8.4\text{-}1)$$

where $E_p(0)$ is the phasor amplitude of the field after the nth round trip, E_{pi} is the incident optical field phasor, n_2 is the nonlinear refractive index coefficient, and where we have assumed the linear refractive index n of the nonlinear sample to be equal to 1. The two exponentials in the second term on the right-hand side of Eq. (8.4-1) depict the phase shifts due to the combined effects of linear and nonlinear propagation, respectively, around the ring cavity.

Possible steady-state solutions to Eq. (8.4-1) are given by setting $E_{p_{j+1}}(0) = E_{p_j}(0) \triangleq E_p(0)$, $E_{p_{j+1}}(L) = E_{p_j}(L) \triangleq E_p(L)$. Substituting this in Eq. (8.4-1), and solving for E_{p0}, we obtain

$$E_p(0) = \frac{T^{1/2}E_{pi}}{1 - R\exp[-jk_0(2L + 2l)]\exp[-jk_0\varkappa_2|E_p(0)|^2L]}.$$

$$(8.4\text{-}2)$$

Now, the output phasor field E_{pt} is given by

$$E_{pt} = T^{1/2}E_p(L)$$

$$= T^{1/2}E_p(0)\exp[-jk_0L]\exp[-jk_0\varkappa_2|E_p(0)|^2L]. \quad (8.4\text{-}3)$$

Combining this with Eq. (8.4-2), we obtain the amplitude transmission function

$$\frac{E_{pt}}{E_{pi}} \propto \frac{T\exp[-jk_0(2L + 2l)]\exp[-jk_0\varkappa_2|E_p(0)|^2L]}{1 - R\exp[-jk_0(2L + 2l)]\exp[-jk_0\varkappa_2|E_p(0)|^2L]},$$

$$(8.4\text{-}4)$$

where we have purposely multiplied the RHS by $\exp[-jk_0(L + 2l)]$. Since we are finally interested in the ratio of intensities, it is clear that the introduction of the extra exponential term will not affect the calculations. Now note that the effective contribution from the phase in the exponents in Eq. (8.4-4) is the same as that from the difference of this phase from the nearest multiple of 2π. Thus, calling

$$\theta = 2k_0(L + l) + k_0\varkappa_2|E_p(0)|^2L - \text{(nearest multiple of }2\pi),$$

$$(8.4\text{-}5)$$

we can recast Eq. (8.4-4) in the form

$$\frac{E_{pt}}{E_{pi}} = \frac{Te^{-j\theta}}{[1 - Re^{-j\theta}]}.$$

$$(8.4\text{-}6)$$

The intensity transmittance becomes, using the relation $R + T = 1$,

$$\frac{I_t}{I_i} = \frac{1}{[1 + 4(R/T^2)\sin^2(\theta/2)]}, \tag{8.4-7}$$

where $I_t = |E_{pt}|^2$ and $I_i = |E_{pi}|^2$. Note that with this definition we can write $|E_p(0)|^2 = |E_p(L)|^2 = |E_{pt}|^2/T = I_t/T$, so that Eq. (8.4-5) becomes

$$\theta = 2k_0(L + l) + \frac{k_0 n_2 I_t L}{T} - \text{(nearest multiple of } 2\pi).$$

$$\triangleq \theta_0 + \theta_2 I_t. \tag{8.4-8}$$

Equation (8.4-7) bears a resemblance to the transmittance of the Fabry–Perot etalon, developed in Chapter 3 [see Eq. (3.2-60)]. Ordinarily, this is plotted as a function of the incident frequency [or δ as in (3.2-60)]. However, in Figure 8.11(a), we have chosen to plot the intensity transmittance as a function of the transmitted intensity I_t (see Eqs. (8.4-7), (8.4-8)), where we also superpose the graphs of I_t/I_i for various values of I_i. The latter graphs are indicated by straight lines A, B, and C. The intersection of these straight lines with the so-called cavity resonance curve gives the steady-state solutions for various values of I_i and can provide a physical explanation of dispersive bistability, as will be seen later.

Note that near a cavity resonance, $|\theta| \ll 1$, so that from Eq. (8.4-7),

$$\frac{I_t}{I_i} = \frac{1}{[1 + (R/T^2)\theta^2]}. \tag{8.4-9}$$

A typical plot of I_t versus I_i is shown in Figure 8.11(b) and clearly shows the cubic variation of I_i with I_t [see Eqs. (8.4-8) and (8.4-9)]. For bistability to occur, we must have a region where $dI_t/dI_i < 0$ or $dI_i/dI_t < 0$. Using Eqs. (8.4-8) and (8.4-9), we can find the boundary of this region by setting $dI_i/dI_t = 0$. A simple calculation shows that the required condition is

$$\theta_2 I_t = -\left(\frac{2}{3}\right)\theta_0 \pm \frac{[\theta_0^2 - 3T^2/R]^{1/2}}{3}, \tag{8.4-10}$$

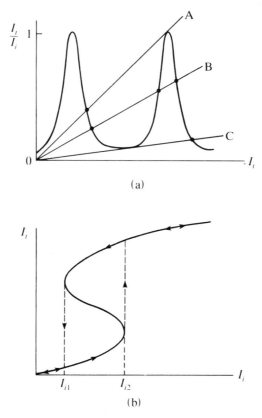

(a)

(b)

Figure 8.11 Steady-state relationships for the purely dispersive case. (a) Transmission peaks of the nonlinear cavity. The straight lines A, B, and C correspond to increasing input intensities. (b) Plot of I_t versus I_i showing the S-shaped curve and possible hysteresis.

where for real solutions, the term in square brackets must be positive.

To intuitively predict the stability of the steady-state solution as plotted in Figure 8.11(b), consider line B in Figure 8.11(a), drawn for a given I_i, which intersects the curve at three points, two of which are to the right hand sides of the two resonance peaks, while one is to the left. A small initial increase of the cavity field (proportional to I_t) for a point on the right hand side [see Figure 8.11(a)] causes the field to eventually decrease (note the variation of

I_t/I_i about that point), whereas an initial decrease in the cavity field causes it to subsequently increase. Thus, such a point is *stable*. The opposite is true for the point in Figure 8.11(a) that lies on the left-hand side of a peak; thus, such a point is *unstable*. From this argument, we note that, for a given value of I_i, the line B in Figure 8.11(a) has two such stable outputs; hence, the system is *bistable*. The range of values of I_i for which there are two stable and one unstable state is marked by I_{i1} and I_{i2} in Figure 8.11(b). For inputs I_i corresponding to lines A and C, there is only one stable output I_t in each case.

Figure 8.11(b) also shows the *hysteresis loop* that I_t follows when I_i is slowly increased and then decreased. The transmitted intensity I_t, initially on the lower stable branch, *switches* to the upper branch as I_i increases past its value for the lower turning point (where $dI_t/dI_i = \infty$) and stays on this upper stable branch for higher values of I_i. As I_i is reduced, I_t still remains on the upper branch until I_i decreases below its value for the upper turning point. At this point, I_t switches back to the lower stable branch. A detailed mathematical treatment of bistability and hysteresis is outside the scope of this discussion.

We remark, before concluding this section, that bistability can be observed in the nonlinear cavity for the purely absorptive case as well. Furthermore, the phenomenon of bistability has been reported from several other optical configurations, namely, during optical transmission across a linear–nonlinear interface [Kaplan (1981) and Cao and Banerjee (1989)] and in multiple quantum-well structures [Gibbs (1985)]. The major difference between the bistable phenomenon studied in this section and the acoustooptic bistability in Chapter 6 is that in the nonlinear Fabry–Perot cavity, the nonlinearity is intrinsic to the medium or cavity. However, in the case of acoustooptic bistability, the nonlinearity is basically from the nonlinear input–output relationship between the sound and the scattered light.

8.5 Phase Conjugation

Optical phase conjugation is a technique that employs the cubic nonlinearity of the optical medium to reverse precisely both the direction and the overall phase factor in an arbitrary beam of light. We can regard the process as reflection of light from a mirror

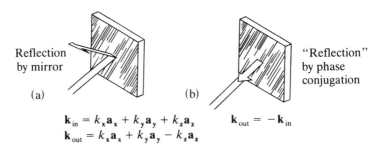

Reflection
by mirror

(a)

"Reflection"
by phase
conjugation

(b)

$$\mathbf{k}_{in} = k_x \mathbf{a}_x + k_y \mathbf{a}_y + k_z \mathbf{a}_z$$
$$\mathbf{k}_{out} = k_x \mathbf{a}_x + k_y \mathbf{a}_y - k_z \mathbf{a}_z$$

$$\mathbf{k}_{out} = -\mathbf{k}_{in}$$

Figure 8.12 Principle of phase conjugation illustrated through a comparison between (a) conventional reflection from a mirror and (b) reflection by phase conjugation.

with unusual image-transformation properties. To illustrate this point, note that a conventional mirror [Figure 8.12(a)] changes the sign of the **k** vector component normal to the mirror surface while leaving the tangential component unchanged. On the other hand, phase conjugation [Figure 8.12(b)] causes an inversion of the entire vector **k**, so that the incident ray exactly returns upon itself [Yariv and Fisher (1983)]. It is not hard to argue that an incident converging (diverging) beam would be conjugated into a diverging (converging) beam. Mathematically speaking, this means that if ψ_p represents the incident phasor, then ψ_p^* would represent the phase-conjugated phasor.

8.5.1 Comparison with Holography

At this time it is instructive to refresh our memory with the basic principles of wavefront–reconstruction imaging, or holography, discussed in Chapter 5. Consider two wavefronts, represented by their respective envelopes,

$$\psi_{e1} = a_1 \exp(-j\phi_1)$$
$$\psi_{e2} = a_2 \exp(-j\phi_2),$$

(8.5-1)

to interfere in space, resulting in an intensity

$$I = |\psi_{e1} + \psi_{e2}|^2$$
$$= |\psi_{e1}|^2 + |\psi_{e2}|^2 + \psi_{e1}\psi_{e2}^* + \psi_{e1}^*\psi_{e2}$$
$$= a_1^2 + a_2^2 + 2a_1 a_2 \cos(\phi_1 - \phi_2).$$

(8.5-2)

The intensity distribution is then recorded on a film and the film developed to generate a hologram with a transparency function that is proportional to I. To understand the reconstruction process, assume that the hologram is now illuminated with a reconstruction wave ψ_{e3}. The field behind the transparency is then proportional to

$$\psi_{e3}\left[|\psi_{e1}|^2 + |\psi_{e2}|^2 + \psi_{e1}\psi_{e2}^* + \psi_{e1}^*\psi_{e2}\right]. \qquad (8.5\text{-}3)$$

Note that if ψ_{e3} is an exact duplication of one of the original waves, namely, ψ_{e2}, which we can call the reference wave, the third term in Eq. (8.5-3) becomes equal to $\psi_{e1}|\psi_{e2}|^2$ and, hence, is a duplicate of ψ_{e1}, which we can call the object wave. However, if ψ_{e3} becomes ψ_{e2}^*, observe that the fourth term of Eq. (8.5-3) (which we shall name ψ_{e4}) is proportional to ψ_{e1}^* and, hence, represents the *conjugate* of the object wavefront.

From the discussion in the previous paragraph, it is clear that we can achieve phase conjugation using holographic techniques, although in the conventional sense, the process is slow and not real-time, owing to the efforts in making the hologram. Note that in the preceding discussion we tacitly talked in terms of envelopes rather than phasors, because reversal of the entire **k** vector is not possible using holographic techniques. The process of introducing ψ_{p1}, ψ_{p2}, and ψ_{p3} all at once in a cubically nonlinear medium to generate ψ_{p4} speeds up the conjugation and is referred to as the *four-wave mixing problem*. The entire **k** vector of ψ_{p4} is the negative of that of ψ_{p1}. We will discuss this at some length in the following subsection.

8.5.2 Semiclassical Analysis
We start from Eq. (8.1-8) with $\beta_2 = 0$,

$$\frac{\partial^2 \psi}{\partial t^2} - v^2 \nabla^2 \psi = \frac{2\beta_3}{3} \frac{\partial^2 \psi^3}{\partial t^2}. \qquad (8.5\text{-}4)$$

Writing

$$\psi(x, y, z, t) = \text{Re}\left[\psi_p(x, y, z, t)\exp(j\omega_0 t)\right]$$

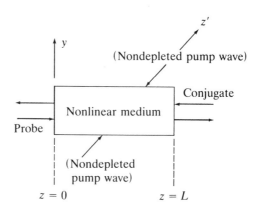

Figure 8.13 Basic geometry of phase conjugation by four-wave mixing.

in Eq. (8.5-4), we obtain

$$2j\omega_0 \frac{\partial \psi_p}{\partial t} - \omega_0^2 \psi_p - v^2 \nabla^2 \psi_p = -\frac{1}{2}\beta_3 \omega_0^2 |\psi_p|^2 \psi_p, \quad (8.5\text{-}5)$$

where we have restricted ourselves around the carrier frequency ω_0 and have assumed ψ_p to be a slowly varying function of time in the sense that

$$\left| \frac{\partial^2 \psi_p}{\partial t^2} \right| \ll \omega_0 \left| \frac{\partial \psi_p}{\partial t} \right|.$$

Consider, now, the geometry of Figure 8.13. We assume the total phasor ψ_p in the interaction region (nonlinear medium) to be the sum of four phasors as

$$\psi_p = \bar{\epsilon}\psi_{p1} + \psi_{p2} + \psi_{p3} + \bar{\epsilon}\psi_{p4} \qquad (8.5\text{-}6a)$$

$$= \bar{\epsilon}\psi_{e1}(t)\exp(-jk_0 z) + \psi_{e2}(t)\exp\left[-j(k_{y0}y + k_{z0}z)\right]$$

$$+ \psi_{e3}(t)\exp\left[j(k_{y0}y + k_{z0}z)\right] + \bar{\epsilon}\psi_{e4}(t)\exp(jk_0 z). \quad (8.5\text{-}6b)$$

In Eq. (8.5-6a), ψ_{p2} and ψ_{p3} represent the phasors corresponding to the *pump* waves, with ψ_{p1} and ψ_{p4} representing the *probe* and the *conjugate* wave phasors, respectively. The quantity $\bar{\epsilon}$ is a bookkeeping parameter which is taken to be small to emphasize that the pump energy is much larger than the probe and conjugate wave energies. Equation (8.5-6b) specifies the direction of propagation of the components of the phasor ψ_p. The pumps are contradirected and travel in directions specified by the direction cosines, $\pm(k_{y0}/\overline{k}_0, k_{z0}/\overline{k}_0)$, respectively, where

$$k_{y0}^2 + k_{z0}^2 = \overline{k}_0^2. \qquad (8.5\text{-}7)$$

The probe and the conjugate travel in the $\pm z$ directions, respectively. Note that we have purposely assumed the propagation constant for the pumps to be different from k_0, the propagation constant of the probe and conjugate. This is because in a nonlinear medium, the effective propagation constant (and hence the effective phase velocity) depends on the amplitudes of waves propagating in the medium. Because the amplitudes of the pumps are much larger than those of the probe and the conjugate, we expect the effective propagation constant to be different in the direction of the pumps, but not in the direction of the probe and conjugate.

Our aim is to study the evolution of the probe and the conjugate waves, as well as to understand the behavior of the pumps. Because all the participating waves propagate in different directions, we find it convenient to track their amplitudes as functions of a common variable. We therefore choose this variable to be t, which has the dimension of time. To be precise, t represents the time each participating wave travels in its specified direction. This is a little different from our approach during the study of second-harmonic generation (see Section 8.2), where we tracked each spectral component as a function of z, because all spectral components travelled in the same direction.

If we substitute Eq. (8.5-6b) into Eq. (8.5-5), the left-hand side yields the following coefficients for the four complex exponentials:

$$\overline{\epsilon}2j\omega_0\frac{d\psi_{e1}}{dt}, \qquad \text{for } \exp(-jk_0z);$$

$$2j\omega_0\frac{d\psi_{e2}}{dt} - \left(\omega_0^2 - v^2\overline{k}_0^2\right)\psi_{e2}, \qquad \text{for } \exp\left[-j(k_{y0}y + k_{z0}z)\right];$$

$$2j\omega_0\frac{d\psi_{e3}}{dt} - \left(\omega_0^2 - v^2\overline{k}_0^2\right)\psi_{e3}, \qquad \text{for } \exp\left[j(k_{y0}y + k_{z0}z)\right];$$

$$\overline{\epsilon}2j\omega_0\frac{d\psi_{e4}}{dt}, \qquad \text{for } \exp(jk_0z);$$

$$(8.5\text{-}8)$$

where we have set $\omega_0/k_0 = v$. Evaluation of the right-hand side is more involved. We find $|\psi_p|^2\psi_p$ and only retain terms containing the exponentials listed in (8.5-8). Furthermore, for the exponentials of the form $\exp(\mp jk_0 z)$ (pertaining to the probe and the conjugate), we only retain terms that involve $\bar{\epsilon}$, neglecting higher powers of $\bar{\epsilon}$. Similarly, for the exponentials of the form $\exp[\mp j(k_{y0}y + k_{z0}z)]$ (pertaining to the contrapropagating pumps), we retain terms involving $\bar{\epsilon}^0$ and neglect higher powers of $\bar{\epsilon}$. We list next the coefficients of the four exponentials in (8.5-8) that represent the contribution from $|\psi_p|^2\psi_p$:

$$\bar{\epsilon}\left[2\psi_{e1}\left(|\psi_{e2}|^2 + |\psi_{e3}|^2\right) + 2\psi_{e2}\psi_{e3}\psi_{e4}^*\right], \quad \text{for } \exp(-jk_0 z);$$

$$\psi_{e2}\left(|\psi_{e2}|^2 + 2|\psi_{e3}|^2\right), \quad \text{for } \exp\left[-j(k_{y0}y + k_{z0}z)\right];$$

$$\psi_{e3}\left(|\psi_{e3}|^2 + 2|\psi_{e2}|^2\right), \quad \text{for } \exp\left[j(k_{y0}y + k_{z0}z)\right];$$

$$\bar{\epsilon}\left[2\psi_{e4}\left(|\psi_{e2}|^2 + |\psi_{e3}|^2\right) + 2\psi_{e2}\psi_{e3}\psi_{e1}^*\right], \quad \text{for } \exp(jk_0 z).$$

$$(8.5\text{-}9)$$

Incorporating (8.5-8) and (8.5-9) into Eq. (8.5-5), we obtain the following set of evolution equations for the probe, conjugate, and pump amplitudes:

$$\frac{d\psi_{e1}}{dt} = j\frac{\omega_0\beta_3}{4}\left[2\psi_{e1}\left(|\psi_{e2}|^2 + |\psi_{e3}|^2\right) + 2\psi_{e2}\psi_{e3}\psi_{e4}^*\right], \qquad (8.5\text{-}10\text{a})$$

$$\frac{d\psi_{e2}}{dt} = j\frac{\omega_0\beta_3}{4}\left[\psi_{e2}\left(|\psi_{e2}|^2 + 2|\psi_{e3}|^2\right)\right] - \frac{j\left(\omega_0^2 - v^2\bar{k}_0^2\right)}{2\omega_0}, \qquad (8.5\text{-}10\text{b})$$

$$\frac{d\psi_{e3}}{dt} = j\frac{\omega_0\beta_3}{4}\left[\psi_{e3}\left(|\psi_{e3}|^2 + 2|\psi_{e2}|^2\right)\right] - \frac{j\left(\omega_0^2 - v^2\bar{k}_0^2\right)}{2\omega_0}, \qquad (8.5\text{-}10\text{c})$$

$$\frac{d\psi_{e4}}{dt} = j\frac{\omega_0\beta_3}{4}\left[2\psi_{e4}\left(|\psi_{e2}|^2 + |\psi_{e3}|^2\right) + 2\psi_{e2}\psi_{e3}\psi_{e1}^*\right]. \qquad (8.5\text{-}10\text{d})$$

Assuming that the pumps are undepleted, we can set $d\psi_{e2}/dt = d\psi_{e3}/dt = 0$. Then, writing

$$|\psi_{e2}| = |\psi_{e3}| = A, \tag{8.5-11}$$

we can evaluate the modified propagation constant $\bar{k}_0$ in the direction of the pumps. From Eq. (8.5-10b) or (8.5-10c), this becomes

$$\bar{k}_0 \simeq k_0\left[1 - \tfrac{3}{4}\beta_3 A^2\right]. \tag{8.5-12}$$

Hence, the field profile along the direction of the pumps becomes

$$\text{Re}\left[A \exp(j\omega_0 t)\exp(-j\bar{k}_0 z') + A \exp(j\omega_0 t)\exp(j\bar{k}_0 z')\right]$$

$$= 2A \cos(\omega_0 t)\cos(\bar{k}_0 z'), \tag{8.5-13}$$

where z' denotes the direction of the pumps and where $\bar{k}_0$ is given by Eq. (8.5-12). Hence, in the direction of the pumps, we observe a standing wave with a spatial period equal to $2\pi/\bar{k}_0$. We can visualize this standing wave to form a phase grating which scatters the probe field and generates the conjugate.

From Eqs. (8.5-10a) and (8.5-10d) with Eq. (8.5-11), we can track the evolution of the probe and the conjugate by means of the coupled ODEs

$$\frac{d\psi_{e1}}{dt} = j\frac{\omega_0\beta_3 A^2}{2}\left[2\psi_{e1} + \psi_{e4}^*\right], \tag{8.5-14a}$$

$$\frac{d\psi_{e4}^*}{dt} = -j\frac{\omega_0\beta_3 A^2}{2}\left[\psi_{e1} + 2\psi_{e4}^*\right], \tag{8.5-14b}$$

where, in writing Eq. (8.5-14b), we have taken the complex conjugate of (8.5-10d). Recall that the variable t in Eqs. (8.5-14a) and (8.5-14b) is representative of the distance travelled by the probe and the conjugate in their respective directions of propagation z and $-z$, respectively. We can thus recast Eqs. (8.5-14a) and (8.5-14b) to ODEs with z as the independent variable provided we replace t by z/v in Eq. (8.5-14a) and by $-z/v$ in Eq. (8.5-14b). With

$\omega_0 = vk_0$, this gives

$$\frac{d\psi_{e1}}{dz} = j\frac{\beta_3 k_0 A^2}{2}(2\psi_{e1} + \psi_{e4}^*), \qquad (8.5\text{-}15\text{a})$$

$$\frac{d\psi_{e4}^*}{dz} = j\frac{\beta_3 k_0 A^2}{2}(\psi_{e1} + 2\psi_{e4}^*). \qquad (8.5\text{-}15\text{b})$$

To solve this system of equations, we introduce

$$\begin{aligned}
\tilde{\psi}_{e1} &= \psi_{e1}\exp\left[-j\beta_3 k_0 A^2 z\right], \\
\tilde{\psi}_{e4} &= \psi_{e4}\exp\left[j\beta_3 k_0 A^2 z\right],
\end{aligned} \qquad (8.5\text{-}16)$$

to recast Eqs. (8.5-15a) and (8.5-15b) in the form

$$\begin{aligned}
\frac{d\tilde{\psi}_{e1}}{dz} &= j\tilde{\delta}\tilde{\psi}_{e4}^*, \\
\frac{d\tilde{\psi}_{e4}^*}{dz} &= -j\tilde{\delta}\tilde{\psi}_{e1},
\end{aligned} \qquad (8.5\text{-}17)$$

where

$$\tilde{\delta} = \frac{\beta_3 k_0 A^2}{2}. \qquad (8.5\text{-}18)$$

The solution to Eq. (8.5-17) for initial conditions $\tilde{\psi}_{e1}(0)$ and $\tilde{\psi}_{e4}(L)$ is given by [Pepper and Yariv (1983)]

$$\begin{aligned}
\tilde{\psi}_{e1}(z) = &-j\left[\frac{|\tilde{\delta}|\sin(|\tilde{\delta}|z)}{\tilde{\delta}\cos(|\tilde{\delta}|L)}\right]\tilde{\psi}_{e4}^*(L) \\
&+ \left[\frac{\cos(|\tilde{\delta}|(z-L))}{\cos(|\tilde{\delta}|L)}\right]\tilde{\psi}_{e1}(0), \qquad (8.5\text{-}19)
\end{aligned}$$

$$\begin{aligned}
\tilde{\psi}_{e4}(z) = &\left[\frac{\cos(|\tilde{\delta}|z)}{\cos(|\tilde{\delta}|L)}\right]\tilde{\psi}_{e4}(L) \\
&+ j\left[\frac{|\tilde{\delta}|\sin(|\tilde{\delta}|(z-L))}{\tilde{\delta}\cos(|\tilde{\delta}|L)}\right]\tilde{\psi}_{e1}^*(0). \qquad (8.5\text{-}20)
\end{aligned}$$

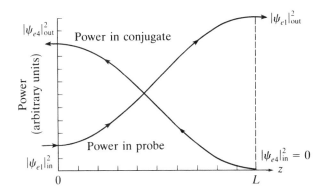

Figure 8.14 Variation of probe and conjugate during propagation through a nonlinear phase-conjugating medium.

In the practical case of phase conjugation, $\tilde{\psi}_{e1}(0)$ is finite, whereas $\tilde{\psi}_{e4}(L) = 0$. In this case, the nonlinearly reflected wave at the input plane ($z = 0$) is

$$\tilde{\psi}_{e4}(0) = -j\left[\frac{\tilde{\delta}}{|\tilde{\delta}|}\tan(|\tilde{\delta}|L)\right]\tilde{\psi}_{e1}^*(0), \qquad (8.5\text{-}21)$$

where $\tilde{\delta}$ is defined in Eq. (8.5-18). Note that $\tilde{\psi}_{e4}$ in Eq. (8.5-21) is proportional to the complex conjugate of $\tilde{\psi}_{e1}$, as expected. The variation of the probe and conjugate power as a function of distance is shown in Figure 8.14.

8.6 The Nonlinear Schrödinger Equation and Soliton Propagation

It is well known in the theory of nonlinear waves that pulses travelling in a nonlinear medium distort under the effect of nonlinearity [Korpel and Banerjee (1984)]. Furthermore, from linear theory, it follows that pulses distort and spread due to dispersion in the medium [Haus (1984)]. This is essentially because different frequency components constituting the pulse travel with different velocities in a dispersive medium [see Section 3.2.4]. It is sometimes possible, in a nonlinear dispersive environment, to ensure distortionless propagation of pulses as a result of a balance between the nonlinearity and dispersion. For baseband propagation, it is a little

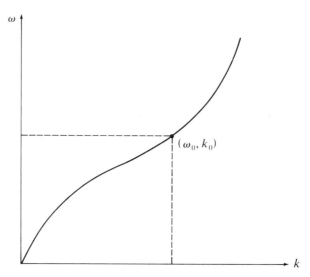

Figure 8.15 Arbitrary dispersion relation showing the variation of the angular frequency ω with the propagation constant k.

easier to see why this should happen: The steepening effect of nonlinearity (see Figure 8.1) can be balanced by the spreading, or smoothening, effect of dispersion. For modulated pulses, both nonlinearity and dispersion cause a chirping in frequency to develop during propagation, as well as other distortions of the envelope. Given the right amounts of nonlinearity and dispersion, it is possible, once again, to offset the effect(s) of one with that of the other to ensure distortionless propagation of the modulated pulse [Korpel and Banerjee (1984)].

In keeping with the nature of the presentation thus far, we will derive the PDE for the envelope in the nonlinear dispersive medium in the fastest possible way. Consider, first, a linear medium with arbitrary dispersion, as shown in Figure 8.15. Because we are discussing the behavior of modulated pulses, it is probably best to locate ourselves at a point (ω_0, k_0) of the dispersion curve (the coordinates corresponding to the carrier frequency and propagation constant) and define *excursions* around that value by a Taylor series expansion of the form

$$\omega - \omega_0 = u_0(k - k_0) + (u_0'/2)(k - k_0)^2, \qquad \omega > 0. \quad (8.6\text{-}1)$$

In Eq. (8.6-1), $u_0 = \partial\omega/\partial k|_{\omega_0}$ and $u_0' = \partial^2\omega/\partial k^2|_{\omega_0}$ are the *group velocity* and *group velocity dispersion* (GVD) terms, respectively. We can calculate these from a knowledge of the explicit equation describing the dispersion curve in Figure 8.15. Now, changing to variables

$$\Delta\omega = \omega - \omega_0, \qquad \Delta K = k - k_0, \qquad (8.6\text{-}2)$$

we can reformulate the nature of the dispersion around the carrier frequency ω_0 as

$$\Delta\omega = u_0\,\Delta K + (u_0'/2)(\Delta K)^2. \qquad (8.6\text{-}3)$$

Our aim now is to find the PDE describing envelope propagation in a dispersive medium described by its dispersion relation. We follow here a recipe that was described in Section 3.2.4: Given a linear PDE describing the propagation of a wave function $\psi(z,t)$, substitution of $\psi \sim \exp[j(\omega t - kz)]$ gives the dispersion relation. For instance, take the linear wave equation

$$\frac{\partial^2\psi}{\partial t^2} - v^2\frac{\partial^2\psi}{\partial z^2} = 0.$$

If we substitute ψ as just mentioned, and equate the coefficients of the exponential, we get $\omega^2 = v^2 k^2$, which is the required dispersion relation. In reality, this manipulation amounted to replacing $\partial/\partial t$ by $j\omega$ and $\partial/\partial z$ by $-jk$, where ω and k are the frequency and propagation constant, respectively. Thus, to find the PDE from the dispersion relation, Eq. (8.6-3), we do exactly the inverse: We replace the frequency variable $\Delta\omega$ by $-j\,\partial/\partial t$ and the propagation constant ΔK by $j\partial/\partial z$. The wavefunction these operators must operate on is, however, the envelope ψ_e of ψ; because we are located around the carrier frequency on the dispersion curve.

Recall that, in Section 3.2.4, we replaced ω and k in the dispersion relation by their respective operators to obtain the PDE for the *real* wavefunction ψ; thus, it was required that the dispersion relation $\omega(k)$ be odd. In this case, $\Delta\omega$ as a function of Δk may not be odd in general because we are interested in describing the evolution of a *complex* wavefunction ψ_e. Using the above recipe,

Eq. (8.6-3) yields the linear PDE [Korpel and Banerjee (1984)]

$$j\left(\frac{\partial \psi_e}{\partial t} + u_0 \frac{\partial \psi_e}{\partial z}\right) - \frac{u_0'}{2} \frac{\partial^2 \psi_e}{\partial z^2} = 0, \qquad (8.6\text{-}4a)$$

or

$$j\frac{\partial \psi_e}{\partial t'} - \frac{u_0'}{2} \frac{\partial^2 \psi_e}{\partial z'^2} = 0 \qquad (8.6\text{-}4b)$$

if we are in a travelling frame of reference, that is,

$$z' = z - u_0 t, \qquad t' = t. \qquad (8.6\text{-}5)$$

In what follows, we modify Eqs. (8.6-4a) and (8.6-4b) by heuristically incorporating the effects of nonlinearity. Assuming a cubic nonlinearity ($\beta_3 \neq 0$; $\beta_2 = 0$) and $\psi(z, t) = \text{Re}\{\psi_e(z, t) \times \exp[j(\omega_0 t - k_0 z)]\}$ in Eq. (8.1-3b), and using the slowly varying envelope approximation, it follows that ψ_e evolves according to

$$j\frac{\partial \psi_e}{\partial t'} + \frac{\omega_0 \beta_3}{4} \psi_e^2 \psi_e^* = 0 \qquad (8.6\text{-}6)$$

in a suitable travelling frame of reference. In our heuristic approach, we include the effects of nonlinearity in Eqs. (8.6-4a) and (8.6-4b) by simply replacing the $j\,\partial \psi_e/\partial t'$ in Eq. (8.6-4b) by $j\,\partial \psi_e/\partial t' + (\omega_0 \beta_3/4)\psi_e^2 \psi_e^*$. This yields

$$j\frac{\partial \psi_e}{\partial t'} - \frac{u_0'}{2} \frac{\partial^2 \psi_e}{\partial z'^2} + \frac{\omega_0 \beta_3}{4} \psi_e^2 \psi_e^* = 0, \qquad (8.6\text{-}7)$$

which is the *nonlinear Schrödinger* (NLS) *equation* describing envelope propagation in a (cubically) nonlinear, dispersive medium [Akhmanov, Sukhorukov, and Khokhlov (1966)]. The equation has been extensively used to model waves in plasma, water waves, electron and ion cyclotron waves, and, with minor changes of the independent variables, pulse propagation through a single-mode nonlinear optical fiber [Agrawal (1989)]. We remark that the nonlinearities in fibers are Kerr-type, whereas the dispersion is essentially material dispersion, because no modal dispersion effects exist.

To find particular solutions of the NLS equation, note that it has the same form as Eq. (8.3-2) that describes beam propagation through a cubically nonlinear medium. In fact, an interesting analogy of pulse propagation in one dimension and beam propagation may be made by comparing Eqs. (8.6-7) and (8.3-2). In fact, we can view diffraction as being similar to dispersion in space. The (real) amplitude of a particular solution is therefore of the same form as Eq. (8.3-9); a plot appears in Figure 8.7(a), with suitably renamed coordinates. These pulses are called *envelope solitons*, and are formed due to a balance between nonlinearity and dispersion. Solitons have been generated and propagated over long distances through optical fibers [Mollenauer and Stolen (1982)], leaving open the door for innovative communication systems that can be much faster than the present ones that suffer from the effects of material dispersion. The advantages of solitons are that they not only remain undistorted by themselves in a nonlinear dispersive medium, but can also interact with each other without a change in shape [Dodd et al (1982)].

Physically speaking, the roles of nonlinearity and GVD in the formation of an envelope soliton may be understood as follows. Consider a modulated Gaussian,

$$\psi = \psi_0 \exp\left[-\left(\frac{t}{t_0}\right)^2\right]\cos \omega_0 t, \qquad z = 0, \qquad (8.6\text{-}8)$$

travelling in a medium described by the dispersion relation Eq. (8.6-1). After a distance L of propagation, different frequency components of the pulse accumulate relative phase shifts due to dispersion because of the different times of travel. For instance, whereas the center frequency ω_0 arrives at $\tau_{\omega_0} = L/u_0$, the sidebands $\omega_0 \pm \Delta\omega$ arrive at

$$\tau_{\omega_0 \pm \Delta\omega} = \frac{L}{u_0 \pm (u_0'/u_0)\Delta\omega} \approx \frac{L}{u_0} \mp \frac{L}{u_0^3}u_0'\,\Delta\omega. \qquad (8.6\text{-}9)$$

Thus, GVD introduces *frequency chirping*; the instantaneous frequency deviation as a function of $\bar{t} = t - z/u_0$ is of the form

$$\Delta\omega = \Delta\omega_d \approx -\left[\frac{u_0^3}{Lu_0'}\right]\bar{t}. \qquad (8.6\text{-}10)$$

On the other hand, the nonlinearity β_3 in the medium also contributes to an extra phase shift of the Gaussian packet due to the amplitude-dependent refractive index. Note, from Eq. (8.1-23), that β_3 is in fact related to n_2 according to $\beta_3 = -n_2/n$. Using this information, the extra phase shift $\Delta\phi$, responsible for what is commonly called *self-phase modulation* (SPM), is

$$\Delta\phi = -n\,n_2 k_0 \psi_0^2 \exp\left[-2\left(\frac{\bar{t}}{t_0}\right)^2\right]\frac{L}{2}$$

$$= n^2 k_0 \beta_3 \psi_0^2 \exp\left[-2\left(\frac{\bar{t}}{t_0}\right)^2\right]\frac{L}{2}. \qquad (8.6\text{-}11)$$

The effect of SPM is frequency chirping as well; in fact, we can find the instantaneous frequency deviation by differentiating Eq. (8.6-11) with respect to $\bar{t}$. The result is

$$\Delta\omega = \Delta\omega_n \approx \left[\frac{n\,n_2 k_0 \psi_0^2 L}{t_0^2}\right]\bar{t}$$

$$= -\left[\frac{n^2\beta_3 k_0 \psi_0^2 L}{t_0^2}\right]\bar{t}, \qquad (8.6\text{-}12)$$

around $\bar{t} = 0$.

It turns out that the distance at which the initial Gaussian pulse packet may evolve into a soliton can be predicted from the nature of frequency chirping due to nonlinearity and dispersion. Note that, for $u_0' > 0$ and $n_2 > 0$, the slopes $s_d = d(\Delta\omega_d)/d\bar{t}$ and $s_n = d(\Delta\omega_n)/d\bar{t}$ have opposite signs and, hence, the effect of one can be balanced by that of the other. In fact, by setting the magnitudes of these slopes equal to each other, we can find the required length $L = L_0$ that the Gaussian packet must propagate to evolve into a soliton. From Eqs. (8.6-10) and (8.6-12), this becomes

$$L_0^2 = \frac{u_0^3 t_0^2}{u_0' n \varkappa_2 k_0 \psi_0^2}. \qquad (8.6\text{-}13)$$

It is important that u_0' and $\varkappa_2$ have the same sign if we expect a soliton to form.

Details of soliton interactions in nonlinear dispersive materials are complicated and will not be discussed here. For computer simulations of Gaussian pulse propagation through dispersive and nonlinear materials, we refer readers to Agrawal (1989).

Problems

8.1   Derive Eq. (8.1-7) starting from Eq. (8.1-3).

8.2   In a quadratically nonlinear medium characterized by the quadratic nonlinearity tensor d_{ijk}, assume that there are two propagating electric fields,

$$E_j^{(\omega_1)}(t) = \text{Re}\left[E_{jp}^{(\omega_1)} e^{j\omega_1 t} \right], \qquad E_k^{(\omega_2)}(t) = \text{Re}\left[E_{kp}^{(\omega_2)} e^{j\omega_2 t} \right].$$

Show that the nonlinear polarization phasor around $\omega_1 + \omega_2$ defined by

$$P_i^{(\omega_1 + \omega_2)} = \text{Re}\left[P_{ip}^{(\omega_1 + \omega_2)} \exp\left(j(\omega_1 + \omega_2)t \right) \right]$$

is given by

$$P_{ip}^{(\omega_1 + \omega_2)} = 2 d_{ijk} E_{jp}^{(\omega_1)} E_{kp}^{(\omega_2)}.$$

[*Hint*: Assume that $d_{ijk} = d_{ikj} = d_{jik} = d_{kji}$.] Repeat the problem for the degenerate case where $\omega_1 = \omega_2 = \omega_0$.

8.3 Verify Eqs. (8.2-35), (8.2-36), and (8.2-37) by starting from Eqs. (8.2-23a), (8.2-23b), and (8.2-24) and solving the coupled set of differential equations with initial conditions $u(z = 0) = \bar{\epsilon}v(0)$, $w(z = 0) = w(0)$, where $\bar{\epsilon}$ is a small number.

8.4 Consider the case of second-harmonic generation in the phase-mismatched case. Find the *exact* solution for the second-harmonic amplitude as a function of propagation (z) by starting with zero initial conditions. How does your result compare with the perturbation analysis results in Eq. (8.2-47)?

8.5 A cubically nonlinear material is subjected to a dc electric field Ψ_0. Show that an effective quadratic nonlinearity is induced in the system. Determine the value of the effective quadratic and cubic nonlinearity coefficients in terms of β_3 and Ψ_0. Physically argue how the application of the dc bias can cause second-harmonic generation in a cubically nonlinear material.

8.6 In the presence of a cubic nonlinearity ($n_2 > 0$), it has been shown that a beam with a sech-type profile represents a nonspreading solution. Perform a perturbation analysis on the nonspreading solution and describe the stability of the sech-type beam.

8.7 In higher dimensions, solutions to the beam profile that tend to zero as r (transverse coordinate) tends to infinity are plotted in Figure 8.8. Numerically solve the ODE Eq. (8.3-13) for other initial values. What can you say in general about the nature of these solutions as $r \to \infty$?

8.8 Starting from the paraxial wave equation in a cubically nonlinear medium [see Eq. (8.3-2)], show that, during the propagation of the complex envelope ψ_e, there is conservation of power. To do this, assume $\psi_e(x, z) = A(x, z)\exp[-j\phi(x, z)]$ and show that

$$\int_{-\infty}^{\infty} A^2(x, z)\, dx = \text{constant}.$$

8.9 Suppose three propagating frequencies ω_0 and $\omega_0 \pm \Omega$ exist in a nonlinear medium modelled by Eq. (3.1-7) with $\beta_2 = 0$. Assuming the amplitudes of the sidebands to be equal, find constant-amplitude propagating plane-wave solutions of the system and the equivalent phase velocity.

8.10 In absorptive optical bistability, the second exponential in Eq. (8.4-1) is purely real and of the form $e^{-\alpha L}$, where α is a small absorption coefficient given by

$$\alpha = \frac{\alpha_0}{1 + I_t/T}.$$

Find and plot I_t versus I_i for various values of $\alpha_0 L/T$.

8.11 Verify Eqs. (8.5-19) and (8.5-20) by starting from Eq. (8.5-17).

8.12 Analyze phase conjugation when the probe (and the conjugate) are modulated pulses. To do this, you must assume the probe and conjugate envelopes to be functions of z and t. Show that Eqs. (8.5-17) are modified to

$$\frac{\partial \tilde{\psi}_{e1}}{\partial z} + \frac{1}{v}\frac{\partial \tilde{\psi}_{e1}}{\partial t} = j\tilde{\delta}\tilde{\psi}_{e4}^*,$$

$$\frac{\partial \tilde{\psi}_{e4}^*}{\partial z} + \frac{1}{v}\frac{\partial \tilde{\psi}_{e4}^*}{\partial t} = -j\tilde{\delta}\tilde{\psi}_{e1}.$$

8.13 A nonlinear dispersive medium is characterized by a nonlinear refractive index coefficient n_2, group velocity u_0, and group velocity dispersion u_0'. Find the exact expression for the envelope soliton that may exist in the medium. Hence find the energy of the soliton.

References
8.1 Abramowitz, M. and I. Stegun (1965). *Handbook of Mathematical Functions*. Dover, New York.

8.2 Agrawal, G. (1989). *Nonlinear Fiber Optics*. Academic, New York.

8.3   Akhmanov, S. A., A. P. Sukhorukov, and R. V. Khokhlov (1966). *Sov. Phys.—JETP* **23** 1025.

8.4   Banerjee, P. P. and A. Korpel (1981). *J. Acoust. Soc. Amer.* **70** 157.

8.5   Banerjee, P. P., A. Korpel, and K. E. Lonngren (1983). *Phys. Fluids* **26** 2393.

8.6   Bloembergen, N. (1965). *Nonlinear optics*. Benjamin, New York.

8.7   Cao, G. and P. P. Banerjee (1989). *J. Opt. Soc. Amer. B.* **6** 191.

8.8   Chiao, R. Y., E. Garmire, and C. H. Townes (1964). *Phys. Rev. Lett.* **13** 479.

8.9   Dodd, R. K., J. C. Eilbeck, J. Gibbon, and H. Morris (1982). *Solitons and Nonlinear Wave Equations*. Academic Press, New York.

8.10   Ghatak, A. and K. Thygarajan (1978). *Contemporary Optics*. Plenum, New York.

8.11   Gibbs, H. M. (1985). *Optical Bistability: Controlling Light with Light*. Academic Press, New York.

8.12   Haus, H. A. (1966). *Appl. Phys. Lett.* **8** 128.

8.13   Haus, H. A. (1984). *Waves and Fields in Optoelectronics*. Prentice-Hall, Englewood Cliffs, NJ.

8.14   Kaplan, A. E. (1981). Theory of plane wave reflection and refraction by the nonlinear interface. In *Optical Bistability* (C. M. Bowden, M. Ciftan, and H. R. Robl, eds.). Plenum, New York.

8.15   Korpel, A. and P. P. Banerjee (1984). *Proc. IEEE* **72** 1109.

8.16   Korpel, A., K. E. Lonngren, P. P. Banerjee, H. K. Sim, and M. R. Chatterjee (1986). *J. Opt. Soc. Amer. B.* **3** 885.

8.17   Lugiato, L. (1984). Theory of optical bistability. In *Progress in Optics*, Vol. 21 (E. Wolf, ed.). North-Holland, Amsterdam.

8.18   Mollenauer, L. F. and R. H. Stolen (1982). *Fiberopt. Technol.* 193.

8.19   Pepper, D. M. and A. Yariv (1983). Optical Phase conjugation using three-wave and four-wave mixing via elastic photon scattering in transport media. In *Optical Phase Conjugation* (R. A. Fisher, ed.). Academic Press, New York.

8.20   Phillips, O. M. (1974). Wave interactions. In *Nonlinear Waves* (S. Leibovich and A. R. Seebass, eds.). Cornell University Press, Ithaca, New York.

8.21 Whitham, G. B. (1974). *Linear and Nonlinear Waves*. Wiley, New York.

8.22 Yariv, A. (1985). *Introduction to Optical Electronics*. Wiley, New York.

8.23 Yariv, A. and R. A. Fisher (1983). In *Optical Phase Conjugation* (R. A. Fisher, ed.), Academic Press, New York.

8.24 Yariv, A. and P. Yeh (1984). *Optical Waves in Crystals*. Wiley, New York.

Index

Index

Index